Spinning Wheels, Spinners and Spinning

SPINNING WHEELS
Spinners and Spinning

Patricia Baines

B T Batsford Limited, London

First published 1977
First paperback edition 1982
Copyright Patricia Baines 1977

Phototypeset by Tradespools Ltd, Frome, Somerset
Printed by The Anchor Press Ltd,
Tiptree, Essex
for the publishers
B. T. Batsford Limited,
4 Fitzhardinge Street,
London W1H 0AH

ISBN 0 7134 0822 7

Contents

List of Illustrations   7

Acknowledgements   9

Preface   11

Chapter One   **Fibres and their preparation**   15
               Flax · Cotton · Wool · Silk

Chapter Two   **Spindle wheels**   40
               Spindle wheels in the East · The Oriental or
               Eastern type of spindle wheel in Europe · Hoop
               rim spindle wheels · Spindle wheels in the
               British Isles · Accelerating the spindle ·
               Winders

Chapter Three **The flyer spinning wheel**   69
               The stock · The driving wheel · The treadle
               · The mother-of-all and maidens · Tensioner
               · Flyer mechanism · The bobbin · Different
               types of drive · The driving band and threading
               · Early spinning wheels with flyer in Europe

Chapter Four   **Accessories to the spinning wheel**   94
               Distaff · Dressing the distaff · Ribbons and
               bands · Waterpots · Greasepots · Seats
               · Bobbin holders · Reels · Swifts

Chapter Five   **Principal types of spinning wheels**   112
               Picardy spinning wheel · Horizontal spinning
               wheel with stock · Horizontal spinning wheel
               with frame · Vertical spinning wheel with base
               · Vertical spinning wheel with frame.

Chapter Six   **Some special types of spinning wheels**   144
               Castle spinning wheels · Two-wheel spinning
               wheels · Double flyer spinning wheels ·

Chapter six continued Boudoir spinning wheels · Table spinning wheels
Girdle spinning wheels · An improved spinning
wheel · Queen Victoria's spinning wheels

Chapter Seven **Spinners and spinning** 175

Chapter Eight **Practical application** 193
Notes for the beginner on using a flyer spinning wheel
· Making a woollen yarn · Making a worsted
yarn · The short draw · The cotton draw ·
A note on silk · Spinning flax · Plying ·
Reeling · Scouring and drying · Notes on
using the spindle wheel · Notes on using the
spindle · Some variations for carding, spinning
wool, and plying · Design and purpose.

Appendix One **Choosing and sorting wool** 229

Appendix Two **Sheep breeds** 233

Source material 240

Books on practical spinning 246

Publications 247

Index 248

List of Illustrations

Plates

Frontispiece Mrs Lane by Paul Sandby
1 Pulling flax in Co. Antrim
2 Retting, Co. Antrim
3 Flax breaker from Germany
4 Beetling, scutching and hackling
5 Cotton bowing
6 Sheep shearing
7 Processes in making woollen cloth
8 Combing and spinning wool, Fifteenth century
9 Wool combs
10 Wool combing: drawing-off
11 Indian Charka
12 Chinese multi-spindle treadle wheel
13 European rimless spinning wheel
14 Woman spinning cotton by Pehr Hilleström
15 Fourteenth century miniature of woman spinning
16 Spinning and carding, from Luttrell Psalter
17 The spinning room, Waldstein woollen mill
18 Great wheel on farmhouse wall, Faeroes
19 Welsh great wheel
20 Irish spinner using a big wheel
21 Detail of a great wheel
22 Women spinning hemp for fishing nets
23 Foldaway Charka
24 Hurdy wheel from Nova Scotia
25 Spinning wheel with flyer and treadle
26 Detail of flyer mechanism
27 Detail of German spinning wheel with flyer drag
28 Bobbin drag on Shetland spinning wheel
29 Detail of silk throwing machine
30 Drawing of spinning wheel, c.1480
31 Young woman spinning, 1513
32 Anna Codde, 1529
33 Scottish vertical spinning wheel
34 Comb distaff from Lapland
35 Spinning from a tow distaff
36 Seventeenth century engraving, 'The Spinner'
37 Spinning and reeling in Co. Down
38 Spinning chair from the Tyrol
39 Free standing clock or wrap reel
40 Reel with clock indicator
41 'Rembrandt's Mother', c.1630
42 Spinning flax using hand-turned disc wheel
43 Spinning wheel from Lower Saxony
44 Spinning wheel from Norway
45 Richard Arkwright's spinning wheel
46 Samuel Crompton's spinning wheel
47 French spinning wheel, 1765
48 French spinning wheel with triangular base
49 Horizontal spinning wheel from the Tyrol
50 Horizontal spinning wheel in use in Bavaria
51 Vertical spinning wheel from the Tyrol
52 Vertical spinning wheel from Denmark
53 Scottish spinning wheel with boomerang supports
54 Vertical spinning wheel from Bavaria
55 Vertical spinning wheel from Switzerland
56 Scottish castle wheel
57 Irish castle wheel in use
58 American chair spinning wheel
59 Two-handed vertical spinning wheel, 1681
60 Flyers on Scottish two-handed spinning wheel
61 Double flyer vertical spinning wheel from Germany
62 Scottish vertical spinning wheel with three flyers
63 Multiple spinning wheel
64 Boudoir spinning wheel
65 Table spinning wheel
66 Girdle spinning wheel
67 Detail of spinning wheel by Doughty of York
68 Spinning wheel by John Planta of Fulneck
69 Spinning wheel by James McCreery of Belfast
70 Queen Victoria at her spinning wheel
71 'The Spin House' from 'Something for All'
72 Cotton workers from Guest's 'History', 1823

LIST OF ILLUSTRATIONS CONTINUED

73 Crank in position to commence Z twist
74 Carding: placing wool on carder
75 Carding: commencing
76 Carding: transferring wool left to right
77 Carding: transferring wool right to left
78 Carding: easing off wool
79 Carding: forming a rolag
80 Woollen spinning: joining
81 Woollen spinning: start of the long draw
82 Woollen spinning: fibres partially
    attenuated
83 Woollen spinning: fibres fully attenuated
84 Combing wool
85 Worsted spinning: drafting
86 Worsted spinning: smoothing
87 Rolling carded wool across carders
88 Drum carder
89 Short draw
90 Cotton draw
91 Flax: preparing the fibres
92 Flax: spinning
93 Plying a uniform yarn
94 Reeling off
95 Joining on a spindle wheel
96 Plying for making fancy yarns
97 Various fibres and yarns
98 Fibres and yarns
99 Detail of coat fabric
100 Lace made with hand-spun linen

Figures

1 Sheep shears
2 Spindles
3 Spindle wheel
4 Z twist; S twist
5 Cross-section of a tensioner
6 Flyer mechanism
7 Roll of flax on bat distaff
8 Lazy Kate
9 Distaff for bobbins
10 Stick reel
11 Hank holder, swift and rice
12 Picardy-type flyer
13 Horizontal spinning wheel from
   Switzerland
14 Rolling flax on to distaff
15 Tying hanks of yarn
16 Method of plying wool from single thread
17 Layout of fleece

Acknowledgements

I would like firstly to express my thanks to members of the staff of all the museums and libraries who have assisted in my search for material for this book by allowing access to stores, opening cases, giving of their time to answering questions, letters and producing photographs. In particular, Schweizerisches Museum für Volkskunde, Basel; Rätisches Museum, Chur; Nationalmuseet, Dansk Folkemuseum, Brede; Norsk Folkemuseum, Oslo; Nordiska Museet, Stockholm; Tiroler Volkskunst-Museum, Innsbruck; Musée National des Arts et Traditions Populaires, Paris; The American Museum in Britain, Bath; Welsh Folk Museum, St Fagans; National Museum of Antiquities of Scotland, Edinburgh; Manx Museum, Isle of Man: and my personal thanks to Dr Ulrike Zischka, Museum für Deutsche Volkskunde, Berlin; H. Tietzel, Deutsches Museum, Munich; J. W. Krom, Stichting Twents-Gelders Textielmuseum, Enschede; G. Hollingshead, Bradford Industrial Museum; P. C. D. Brears, Castle Museum, York; J. C. S. Magson, Bankfield Museum, Halifax; C. Gilbert, Temple Newsam House, nr. Leeds; A. J. de la Mare, Snowshill Manor, Broadway; A. Montgomery, Ulster Museum, Belfast; Mrs Laura Jones, Ulster Folk and Transport Museum, Holywood; and Robert Miles, Science Museum, London who encouraged me from the start and has answered so many questions.

Dr Ilid E. Anthony, Welsh Folk Museum, I wish to thank for information about flax growing in Wales, page 137, and Mrs Grace R. Cooper, Division of Textiles, Smithsonian Institution for information about Amos Minor, page 66.

To my numerous spinner friends who have given me help and advice in so many ways, I give my most sincere thanks; in particular to members of the Oxford Guild of Weavers, Spinners and Dyers, and to my teachers, Gerald Carter who first led me into the realms of spinning, the late Constance Towers who taught me to spin flax, and Morfudd Roberts to whom I owe much valuable practical knowledge. I also thank David Lane for making such a fine job of the detail photographs showing techniques in chapter VIII.

And finally to my husband, who has helped me beyond measure with patience and encouragement through all the stages of this book including checking the manuscript and finding time to do the line drawings, my gratitude is expressed in the dedication.

To Tony

Turn the reel,
Spin the wheel,
Spin – spin.
Wind it full
With finest wool,
That every lad
May wear his plaid,
Turn the reel,
Spin the wheel,
Spin – spin.

ANON.

Preface

In writing this book I have attempted to give some indication of the evolution of the spinning wheel, some of the differences in its design, something of the people who used them, and some of the techniques that have been used. The history of the spinning wheel is closely linked to the various textile industries in which its efficiency as a work tool was of first importance. However, it has been no less valuable in domestic life, and both spheres have contributed to its evolution, the two often interdependent.

For the part it has played in the home the spinning wheel is of interest to the folklorist, but according to its finished appearance it was as acceptable in royal households as in the humblest cottage. When made in a refined manner and of good quality materials it becomes of interest to the furniture collector, while to those interested in its mechanics it is seen as one of the main precursors of the textile industrialisation. For such reasons spinning wheels can be found in all kinds of museums, from the Industrial or Science, to the Folk or Everyday Life; from the Textile to the Historical or Town museums, and sometimes the historical mansion or the stately home. To visit a museum, however small and wherever it may be, might bring the reward of finding some sort of spinning wheel; this is particularly so with the revived interest in crafts of all kinds. However, difficulties arise since many museums lack the space to display everything, but also, unfortunately, have few complete records about the spinning wheels in their possession. Without such background information, making judgements as to their age, their provenance and their specific use can too easily lead to misconceptions. Added to which, spinning wheels are sometimes wrongly assembled, a vital part may be missing or driving bands wrongly linked. They may have been acquired in such a condition, and while this can be of interest, it is misleading if the components are not, as far as possible, assembled correctly.

Whereas types of construction became a tradition and were adhered to from one century to the next, styles of turnery did not necessarily follow that of furniture, so these cannot be relied upon as a method of dating or placing spinning wheels, although they can certainly sometimes be a guide. Fashion of course has played its part, and from such accurate information as can be found, and the occasional dated spinning wheel, one can assume that the majority of spinning wheels still in existence date from not earlier than the second half of the eighteenth century and mostly from the nineteenth century. Driving wheels and bases may be older than other parts which are more vulnerable to breakage.

As for the makers themselves, the marks, initials and names that some of them put on their work give only in some instances a clue as to who they were or when they lived. Nor can we assume that every spinning wheel was made entirely in one workshop, though in many cases the local turner who supplied households with furniture and wooden utensils, was also the maker of spinning wheels. He could go to the blacksmith for the metal spindles and axles, to the cobbler for the leather bearings and thonging. He would no doubt have learnt to make good spinning wheels, not by his own experience as a spinner but through the comments and criticisms of his customers. A spinner has, one could say, a personal relationship with her wheel; she gets to know its idiosyncrasies and how to manage them, and what suits one spinner may not suit another. It can therefore be unwise to form a hasty opinion on the capabilities of an old spinning wheel, not knowing who used it in the past, the fibre used, or the type of thread that was made.

Textile terminology can present problems, since with industrialisation the same words are used as before but now with a more precise meaning and may describe only one part of a process or have a slightly different meaning in a different branch of the industry (see the Textile Institute's *Textile Terms and Definitions*, 1963). With the spinning wheel itself, it is not the quaint words used to describe some of the parts which can lead to misunderstanding, but the use of different terms to denote the same thing. As an example with the types of drive; some people think in terms of bobbin or flyer 'lead' while I personally find it better descriptive to speak in terms of 'drag'. Confusion can sometimes also arise between different types and shapes of spinning wheel and in an attempt to make this clear I have arranged my sections according to what I feel are the basic designs, though very well aware that the variety of detail is enormous – only some of which can be mentioned in this book. Moreover, whatever tidy divisions are made amongst spinning wheels and to whatever generalisations, exceptions can nearly always be found.

An early impression I gained in our English museums was the number of spinning wheels said not to be English, and this led to several trips abroad looking in museums for spinning wheels. The visits were confined to Western Europe and many were far too brief, particularly in Scandinavia where it was only possible to glimpse at the wealth of documented material. While I mention places where I have seen a particular characteristic, this of course does not mean that it cannot be found elsewhere, but only that I myself have not come across it elsewhere. I have given the wheel diameters as near as I know them and whenever possible as a guide to the size of spinning wheel except when they are pictured in use and then the scale is obvious.

Unfortunately I have not been able to visit the United States nor have I been to Canada, so only a little has been included about North American spinning wheels. However, their initial sources must have been European, though American and Canadian spinning wheel makers also had individual ideas.

All students of the spinning wheel will always be indebted to John Horner, who was a textile engineer of flax machinery, an accomplished linguist and who travelled widely on the continent as an adviser on the installation of textile machinery. During his travels early in this century he formed a collection of spinning wheels from many parts of the world and this he presented to the Ulster Museum in Belfast. He also made a study of the linen industry. His book *The Linen Trade of Europe during the Spinning Wheel Period* produced in 1920, one year after his death, has, like his collection (known to so many through the line drawings in the catalogue compiled by G. B. Thompson), been a valuable contribution to our knowledge of spinning wheels.

A wish to spin may come initially from an interest in spinning wheels, or perhaps in a particular one, inherited, bought, or seen in a museum, while for others it is the spinning itself which comes first and leads to an interest in the tools used in making textiles. For the benefit of those who cannot spin, I include the last chapter, which gives some techniques for spinning yarn, largely using flyer spinning wheels. It is, of course, difficult to describe in words and still pictures something which is a continuous movement, a rhythm and co-ordination between hands, foot and fibre, and which also sharpens the sense of feel. Spinning is so much a matter of practice, together with learning to understand the behaviour of fibres and the workings of the spinning wheel. I hope that the inclusion of some alternative methods will not confuse the complete beginner. They are there partly because what may work for one person is not necessarily the best for another, and partly for those who already have a knowledge of spinning but wish to expand this into different techniques or into ways with different fibres and thus discover further the creative possibilities.

Since the availability of new spinning wheels is continually changing it seems inadvisable to make recommendations here. Older spinning wheels found in antique shops are seldom in complete working order, and so should be carefully examined and if possible tried out. For practical use they are often not worth their high price, unless a restorer is to hand or you can restore yourself. Contact with local guilds of weavers and spinners should not be difficult; they exist in most parts of Britain, and in areas of Australia, Canada, New Zealand and the United States, and there are also craft magazines, listed at the back of this book, which advertise equipment and raw material. The craft is once again a flourishing one!

Patricia Baines
Oxford, 1976

A drawing by Paul Sandby c. 1760 of Mrs
Lane using a girdle spinning wheel.
*Photograph: Royal Library, Windsor Castle.
Reproduced by gracious permission of Her
Majesty the Queen*

Chapter One

Fibres and their preparation

Spinning is the technique of twisting together a number of fibres, which can vary in length from $\frac{3}{8}$in. (approx. 1cm.) of cotton fibres to 7ft. (214cm.) of jute, into a strong continuous thread. If a bunch of any textile fibres is held in one hand and with the other a few fibres are drawn-out, these will part company from the bunch, but if the hand drawing-out the fibres at the same time twists them in one direction only, they start to form a thread. Give them more twist and the thread becomes stronger, and continue to draw-out fibres while twisting them and they become a continuous length. Let go this thread and it will immediately untwist, but if wound on to a stick it remains a thread.

When this was first discovered is lost deep in the mists of man's early history, but it is amazing when one stops to consider that up to the middle of the eighteenth century all yarn that was used in the making of fabric, from coarse sacking to the finest lace trimmings, or heavy horse blankets to the lightest shawl, was spun – with the exception of silk filament – by the hands of an individual with no tools other than a spindle and distaff, or a spinning wheel, using materials provided by nature from animals and plants.

As fabric is dependent on yarn, it is important that the yarn, both by the properties of the raw material and by the construction, be capable of withstanding the rigours put upon it by the many processes and the tools used to transform it from fibres into a fabric suitable for its ultimate use.

The concern of this book is with the first part of this process, namely with making fibres into yarn, yet a very brief and basic resumé of what happens to yarn subsequently may be helpful for those unfamiliar with the making of cloth. The processes are numerous and vary according to the fibre and the type of cloth required.

Cloth is woven on a loom, of which there are many different types, by interlacing threads at right angles to each other. The threads that run lengthwise are individually known as ends and collectively the warp. They need to have sufficient strength, and therefore are usually well twisted, to take the tension which is put upon them during the process of weaving, when they are kept taut. It was the usual practice of weavers to size the warp in order to prevent the yarn from fraying as it was rolled from the back of the loom to the front, passing through string or wire eyelets (heddles) slung between two wooden bars (shafts), and through the narrow slots in a metal or cane frame (the reed) to keep it evenly spaced. To size wool, the warp was dipped in a trough of light glue made from boiled white leather or shreds of parchment and then wrung out by passing through a terracotta neck attached to one end of the trough. Linen was sized with starch, such as a flour and water paste which was brushed on to the warp, or it could be rubbed with green soft soap or even tallow.

The two or more shafts are tied underneath the loom to pedals and when one of these is pressed it lowers (or with the use of pulleys can raise) the shaft which separates these pre-arranged warp ends from the rest to form an opening, at a point between the reed and the front cross-bar of the loom, known as the shed. This puts a further strain on the warp. The yarn which interlaces with the warp is called the weft and is placed in a shuttle and thrown by the weaver through the shed across the full width of the warp. The weft can be spun with less twist, and therefore be softer, since it is not subjected to so much strain and needs to bed into the warp as it is beaten into place – a second function of the reed – lightly or hard depending on the character of the cloth. The shed is then changed by depressing a different pedal which lowers (or raises) another set of warp ends. By repeating on these lines cloth is formed.

While the warp is measured to the full length of the cloth, either round posts or on a rotating warping mill, the weft is wound on to hollow tubes, bobbins, spools, pirns or quills, whichever word you care to use (they can be made of wood, rush, cane, goose feathers or stiff paper), by means of a winder. With this, the bobbin is pushed on to a slightly tapered spindle, usually made of metal or wood, turned by a driving wheel. For centuries a bobbin winder stood beside the loom and it has often been mistaken for a spinning wheel. However, to one side of the winder a hank or skein of yarn is placed on a swift, a slightly cone-shaped frame which rotates as the yarn is wound off and guided on to the bobbin by one hand held close to it, the other hand turning the wheel.

Further strength to the yarn can be given by twisting two or more threads together, sometimes called folding, doubling, twining, thickening, but more usually plying, and it must be noted that this can be achieved with the same types of equipment as used for actually spinning.

Colouring by dyeing can be done either on to the yarn, the woven cloth, or, with some fibres, before spinning. To take up the dye the material must be quite clean, and for pale colours may need to be whitened (bleached). Colouring by printing is, of course, on to the cloth. After weaving there are the finishing processes; these include mending broken ends and washing (scouring), but from then on the processes depend on the sort of cloth, its fibre, and the purpose to which it will be put. Some types of woollen cloth are felted by pummelling in water, known as fulling or milling; then put out to dry, stretched between hooks attached to posts, known as tentering; raised to form a hairy pile by scratching up the surface with teasles; the pile then cropped with large heavy-bladed scissors known as croppers, to give the cloth a smooth finish, and finally pressed.

Certain cloths are not milled but the process of pressing, known as calendering, also gives a smooth finish.

The bleaching of linen used to be a lengthy process of scouring it in lye (alkaline – in Ireland fern ash and soap were sometimes used) and laying it out on grass to expose it to moisture, light and air (grassing) for several days. The process was repeated and could take from one to three months to complete. Steeping in weak solutions of chloride of lime is another method of bleaching. Beetling is beating the cloth while damp with wooden or metal hammers to produce a closed lustrous smooth surface.

Yarn is also needed for non-woven fabric: interlooping on needles for knitting and with the use of a hook for crochet and for constructional fabric such as lace. In lace-making, the thread lies between a pin and a small bobbin supported by a pillow and the tension is manipulated by the lace-maker herself, so the thread can be of the very finest. And again, yarn is used, with a single shuttle, for tatting and netting. For the latter, the yarn may need to be tough to endure the hard treatment of its use. Also, of course, thread is used for sewing. It was often the custom, in the days when it was spun by hand, to pass the thread through beeswax to make it stiff and smooth.

The qualities required of a textile fibre are flexibility, strength and a certain amount of elasticity. Of the three groups of natural fibres, those from plants are composed mainly of cellulose while those from animals are based on protein: asbestos is the only mineral fibre.

The fibres from plants, or vegetable fibres, are sub-divided into three groups.

(1) The bast fibres, that is, those that come from beneath the surface of the stem, principally flax, hemp, jute, ramie and nettle.
(2) Leaf fibres, principally sisal and abaca (manila).
(3) The seed fibres, principally cotton.

The fibres from animals can also be divided into three.
(1) Wool, which comes from sheep.
(2) Hair, closely related to wool; some animals produce both but the wool should not be confused with that of sheep. Hair comes from camel, goat, the Angora goat which produces mohair, the Cashmere goat, llama, alpaca, vicuña, rabbits (usually Angora), dogs, cats, and various species of cow, buffalo and ox.
(3) Silk is a continuous filament produced by the silk worm while making its cocoon, but there is silk waste, which is in short lengths and can be spun like the other fibres.

It must be stressed, however, that within each of these types, the fibres vary enormously not only in length but also in quality and strength. These differences are caused by nature herself; climatic conditions, soil conditions and situation each have an influence on the growth of the animal or plant that bears the fibres (a sick sheep, for instance, will produce a weakness in its wool). Through scientific research and the examination of natural fibres, man-made ones (natural polymer and synthetic) have been produced which are reliably uniform and ideal for the total mechanisation of the spinning process in the present day.

However, the natural fibres are the ones that concern us most in connection with the spinning wheel. Flax, cotton, wool, and silk waste, all have been used for many thousands of years for making textiles. Flax and wool are mentioned in the Old Testament and both are indigenous to Europe. Cotton has an ancient history in countries as far apart as India and Peru, in both of which people spun and wove with immense skill, but their species of cotton were different (E. Anderson, 1971). Sericulture and the production of silk cloth had been a practice in China since c.3000 B.C. where for thousands of years it remained a guarded secret.

To trace their early history is not our purpose here, but the latter two, cotton and silk, although not indigenous to Europe had, by the Middle Ages, become part of its textile scene. Certainly the tools used in their production played a considerable role in the development of textile manufacturing processes.

Although in earliest times fibres were probably prepared with the fingers and rudimentary implements, subsequent tools were, in the main, quite simple. Those known to us in Europe from late medieval times remained virtually unchanged until they were mechanised at the time of the Industrial Revolution. The preparation of the raw material is, and always has been inseparable from spinning, and to produce a fine yarn it is first necessary to have good quality raw material and well-prepared fibres.

FLAX

Flax, *Linum usitatissimum*, is cultivated both for its fibres and for its seed which forms the basis of linseed oil. The fibres are formed in bundles which lie in a ring just beneath the outer bark and surround the woody core. Each fibre is made up of short cells which overlap and are held together by a gummy substance, thus forming continuous strands the full length of the plant stem. Under the microscope the fibre is long, smooth, and cylindrical, but has slight bumps on its surface where the cells overlap. These catch on one another during the spinning process and along with the gummy matter help to hold the thread together. Since the bulk of the cell walls have a left-handed spiral the fibres tend to lie in this direction, and, when drying after wetting, follow this spiral. Flax is a relatively non-elastic fibre which is lustrous and makes a strong linen thread which can be spun extremely fine. It is usual to dampen the fibres to soften the gummy matter especially when spinning fine.

Hemp, *Cannabis sativa,* is usually rather coarser and longer than flax and is structurally different, since the outer fibrils spiral in a right-handed direction. The lengthy process of preparation prior to spinning is much the same for both plants.

The wild form of flax, *Linum angustifolium*, is a common weed in the Mediterranean area and it seems unknown whether this derived from the cultivated type or *vice versa*. The wild type has limited distribution in Britain; its five-petalled flowers are a pale blue or white and it produces a very coarse fibre but reasonable linseed oil. The cultivated type is not indigenous to Britain and it was usual to import fresh seed. When grown for oil the plants are placed further apart than when grown for fibre, for which the seeds are sown very close together so that the stems grow straight up, only branching at the top. Flax is an adaptable plant and is grown in a number of countries under different climatic conditions. The seeds are sown in late spring, and growing time is about three months. The plant best for fibre grows to a height of between three and four feet (91–122cm.) and produces a delicate purplish-blue flower. Harvesting takes place as the bottoms of the stems turn yellow. To preserve the maximum fibre length flax is pulled up rather than cut, and for many centuries this was done by hand (1). It can be pulled while it is still green but the seeds will then not be ripe, although the fibres are thought to be at their best. If the crop is harvested late when the seeds are fully mature, more fibre is obtained but the quality is inferior and coarser.

Once harvested, the flax is either stacked or laid out in bundles to dry. The seeds are removed by passing the heads through a rippler, a coarse comb with one line of metal or wooden teeth; this was sometimes fastened in a vertical position to a post in a barn, or attached to a portable bench and the work done in the fields. In Holland and Sweden the rippler was attached to the centre of a bench and used by two people, one sitting at either end, and working together in a rhythm so that they used the rippler alternately.

To reach the fibres themselves it is necessary to rot or ret the straw by exposing it to water, which decomposes the woody matter and cellular tissue holding the fibres together, so as to free them. For this, the flax was made up into bundles called beets or sheaves and immersed in specially dammed rivers, ponds or pits dug in the ground (2). The water, warmed by the sun, needed to be soft and clear; the process of retting in this manner took about seven to ten days. 'In the simplicity of former times when families in England provided themselves with most of the necessaries and conveniences of life, every garden was supplied with a proper quantity of hemp and flax; but the steeping which was necessary to separate the threads, was in many places, found to render the water so offensive and detrimental, that in the reign of Henry VIII a law was made that "No person shall water hemp or flax in any river etc . . . where beasts are used to be watered, on pain of forfeiting, for every time so doing, 20 shilling."' So runs a passage in *Floral Fancies and Morals from Flowers*, written in 1843. In Belgium the River Lys was known as the Golden River because the water was so excellent for retting, but this is now forbidden since the river has become polluted.*

*A more modern method utilises specially-built concrete tanks into which the sheaves are tightly packed. These are usually given a preliminary soaking in cold water to remove the dirt before being soaked in warm water of 30°C for a further three days or so. High quality flax may be double-retted by first putting it partly through the process, drying it and then retting again. (Nowadays chemical methods are also used.)

1 Pulling and binding flax in Co. Antrim,
Northern Ireland c.1910.
*Photograph: Green collection, Ulster Folk and
Transport Museum*

Another method of retting was to spread the crop on the ground, exposing it to dew and rain, though also to sun, so that it needed frequent turning to prevent it from getting too dry on one side and over-rotted on the other, and it even needed to be watered if the weather was dry. Dew-retting can take anything from ten days to three weeks according to conditions. When the tips of the flax are pinched and the fibres splay out and curl, the flax is sufficiently retted. In the *Statistical Accounts of Scotland*, 1793, for the Parish of Wick it is said: '. . . Flax thrives well here, and most of the farmers sow as much as is necessary for the use of their families. . . . Watering it, is found to be a very precarious and troublesome process, requiring a constant regular attention; and taking up much of the farmer's time from other work. The seed is seldom or never preserved here. . . . The experiment of boiling the flax instead of watering it, seems to answer in this Parish.'*

Under-retting makes it difficult to remove the woody matter and separate the bundles of fibres, while over-retting results in beginning to break up the fibres themselves and creates far more waste than is necessary. Dew-retted flax is inclined to be a little coarser than water-retted and darker in colour. The colour of flax varies considerably from a pale white-yellow through various shades of

*I have found that boiling dried flax for 12 hours and leaving it in its water for a further 12 hours is a satisfactory short cut in this tedious process when small quantities are being processed at home.

2 Removing flax from the 'lint hole' after
retting, Co. Antrim.
*Photograph: Green collection, Ulster Folk and
Transport Museum*

fawn and brown to a dark bluish grey. The Courtrai district of Belgium is well
known for its fine flax, as is northern France, the Netherlands and Ireland, all of
which have been famous for the production of fine linen for many centuries.
Richard Hall (*Observations*, 1724) recorded in the eighteenth century that
Ireland had white or yellow flax while Holland and Germany had silver-blue
colours, which became whiter when bleached.

After water-retting the sheaves are stooked to drain off, then untied and the
straw spread out on the ground to dry (also called grassing). Once completely
dried, the flax can be stored until required for the next stages of breaking,
scutching and hackling, known as flax dressing. Before breaking the straw is
usually warmed to get it really brittle. This could be done over an open fire as in
Ireland, but it was a hazardous business and the smoke caused discolouration.
In Holland and Germany ovens were built for the purpose. In its simplest form,
breaking is done by hitting the straw with wooden mallets (in Portugal these
were cylinders made of cork, of which they have so much), but in the fourteenth
century a flax breaker was invented, probably in Holland, and this was used all
over the continent of Europe. The breaker consists of two or three lengths of

3 Decorated flax breaker from Rhön, north-
east of Frankfurt, Germany.
*Photograph: Museum für Deutsche Volkskunde,
Berlin*

wood mounted on their sides on a stand, fixed parallel with each other leaving a
small gap between each. Their upper edges are fined down to form blades.
Above them, one or two corresponding blades, in alignment with the gaps, are
loosely pivoted to the stand at one end and are joined at the other end to make a
handle, resembling a chopper (3). In Denmark the breaking edges of the blades
were sometimes cut in zig-zags and occasionally breakers were built to be used
by two people simultaneously. There seems to be little trace of flax breakers in
England or even Ireland, though in the *Statistical Survey of Co. Monaghan*
(C. Coote, 1801) such a tool is described with the warning, 'Pounding flax is,
evidently, very hurtful; as the blow comes across the grain and must often cut it,
which when it comes to the hackle, will, of course fall into, and waste a considerable quantity of the best part of the flax.' Throughout the three processes of
breaking, scutching and hackling it is important that the root ends are kept
evenly together.

In Ireland there was a method of bruising the flax in the fields, the flax being

4 After plate IV (1791) of a series of 12
engravings illustrating the processes in the
eighteenth-century Northern Ireland linen
industry, by William Hincks. It shows beetling
or flax breaking with a mallet, scutching and
hackling

laid out on the ground in a circle. A large stone wheel attached to a long pole was
anchored by a piece of rope to a heavy stone placed in the centre of the flax
circle to act as a pivot. A horse pulled the wheel round so that it rolled over the
flax, this being turned and moved by the workers so that the wheel passed over
and bruised different parts of the flax each time it came round.

Scutching serves to remove the now broken stalks, called boon (in Ireland,
shoves or shous), by beating the flax with a wooden bat held in the right hand,
against and down a vertical wooden board; there is sometimes a slot through
which the bundles of flax are placed, held and turned by the left hand. The
boards (also used in Ireland) were made to stand upright and the bats could vary
in shape, being often more like sword blades, as shown in Hincks's engraving (**4**).
Very roughly made tools can be found, but in Germany and other parts of
Europe, being an essential part of domestic life, they were wedding presents and
as such were well-finished and decorated. Scutching is sometimes called swing-
ling, and in Germany the bat is called a *Schwingelblatt* and can be somewhat
heart-shaped.

These two operations were mechanised early in the eighteenth century. The
breaking was achieved by passing a handful of flax between fluted rollers, and
the scutching by replacing the bat with a wheel of four or twelve wooden blades.
Hincks, in plate V of his series, shows both operations being turned by a water-
wheel though fed by hand.

To complete the preparation, the flax is passed through combs known as hackles. These consist of circular or square clusters of metal teeth set upright in a wooden board. There are usually three or four different sizes of hackles, each becoming successively shorter and finer. The flax is combed through each set starting with the coarsest and working through to the finest, thus cleaning out the remaining pieces or boon, removing the short fibres and aligning the long ones parallel. Hincks's picture (4) shows the hackles attached to a bench and the work done by a man. This must have been usual in eighteenth-century Ireland, since Richard Hall regretted that the work was not done by women as it was in Holland where, he also notes, hackles were much finer than those made in Coventry. In Holland and Germany the hackle boards were placed in stands. In many European countries these boards usually had a hole at either end so that they could be tied to something firm while in use. There is a delightful painting of 1798 by the Swedish artist, Pehr Hilleström, portraying a woman in a small room in a town house with a makeshift arrangement: a chair turned over on its side with the hackle board tied across it. In the Trentino area of Northern Italy (where all the tools described above were used) and also in Spain in Catalonia, the hackle boards had a semi-circular slot at one end to take the hackler's foot; the board leant against the leg and the flax was drawn upwards towards the body. This position was also used in Northern Ireland but with the board held between the knees and not by the foot. When hackling is completed the flax is stored in bundles known as stricks, and the ends either loosely tied or twisted together.

The long flax is generally known as line, but terms vary and in Scotland it is called lint. There is also short flax, while the short fibres left in the hackles are known as tow. In John Forrester's account book (in the Scottish Record Office) of 1748–9, the headings are lint, breards and tow, breards being the short flax recovered from the first tow by a second hackling.

According to the *Statistical Survey of Co. Monaghan*, '. . . the finest and longest flax is the prime part, the remainder is the backings or the tow, and the refuse of all is used for thatch for houses, for which it is most excellent and durable . . . the flax should be drawn through these [the hackles] several times with great care. The most experienced hands only attempt to work the finest hackle, which is very close and a nice matter to perform well'.

On the subject of storing flax, Andrew Yarraton, an English gentleman who in 1677 wrote a pamphlet entitled *England's Improvements by Sea and Land to out-do the Dutch without Fighting . . .*, has this to say: '. . . there is much in preparing and fitting of the flax, so as to make it run to a fine thread. This is the way they do it in Germany, and thou mayest write by their copy. Thou must twice a year beat thy flax well and dress it well, and take out of it all the filth, and so for as long as thou hast it in thy possession, if it be ten years, and the longer thou keepest it, the finer it will be, for beating and often dressing will cause the Harle [fibre] to open, and at last it will be strangely fine. There must also be a stove in the room where the flax is, with fire in it in all moist times, which is another great cause which makes it so fine. I have seen flax in Saxony twenty years old thus hous-wife't, which was as fine as the hairs of one's head. It is true there what the old saying is here, that wooll may be kept to dirt, and flax to silk.'

COTTON

Cotton comes from the *Gossypium* family of plants and produces its textile fibre inside the seed boll. This fibre first grows lengthwise, distended with liquid, then strengthens its internal structure by adding layers of cellulose; thus the innermost layer is the youngest growth. Each layer is in fact two, one solid and the other porous. As the boll opens the fibres dry out, the cell walls collapse into different shapes of flatness, and because they have been compressed during growth, the fibres twist to form convolutions both to the left and to the right, in an almost equal number on each fibre. These convolutions make it possible for the fibres to adhere to one another during spinning in spite of their smooth surface. Although it is a fairly non-elastic fibre, it has an affinity to wool in that it traps air between the fibres when being spun, though to a lesser degree. Good quality cotton has some natural lustre, but this can be intensified by the process of mercerisation, when the fibres are swollen by a strong solution of caustic soda and dried under tension to prevent shrinkage. The shorter the fibre or staple the coarser it is. The three main division are: short staple, $\frac{3}{8}$–1in. (1–2·5cm.), which is low grade; medium staple, $\frac{1}{2}$–1$\frac{1}{4}$in. (1·3–3·2cm.) which has medium strength and lustre; long staple, 1$\frac{1}{4}$–2$\frac{1}{2}$in. (3·2–6·5cm.), which is lustrous and fine, and includes Egyptian and Sea Island cotton.

Cotton thrives under sub-tropical conditions and needs plenty of rain and sun during its six months growth, and a dry period to mature. The main hazard to the cotton grower is damage through insect-pests that attack the plant during its growth. Although there exists a perennial cotton plant growing to a height of 15–20ft. (456–610cm.) and yielding spinnable cotton for about ten years, the major part of the cotton crop comes from an annual shrub which grows to 1$\frac{1}{2}$–4$\frac{1}{2}$ft. (45·5–137cm.) high. The flower is either yellow or yellow tinged with red at the base, and blooms for only two or three days, after which the seed boll takes a further seven weeks to mature. The ripe bolls are picked, a tedious job when done by hand since they do not all ripen at the same time and most fields have to be picked two or three times before the crop is fully gathered. Unripe bolls will have immature fibres which will be unsuitable for spinning and cause wastage. Such fibres easily get entangled, forming neps which can show in the spun yarn, do not dye easily and appear in the finished fabric a lighter shade.

The next process, known as cotton ginning, is to remove the fibres from the seeds. In India a hand-mill was used, consisting of two rollers, one having five or six horizontal grooves; when turned by a handle in much the same way as a washing mangle, the two rollers rotated in opposite directions. The cotton was fed into one side, and the seeds, being too large to pass between the rollers, were torn off and dropped to the floor. The disadvantage of this slow-moving machine was that many of the seeds did in fact get through, were crushed by the rollers and stained the cotton, greatly reducing its value. The cotton-producing States of America used this same unsatisfactory implement for ginning their cotton until in 1792 an American New Englander from Massachusetts, Eli Whitney, on a visit to friends in Savannah, Georgia, became fascinated by the problems of ginning and invented the machine called a saw gin. Inspired by watching a cat trying to catch chickens through a wire fence, the principle of the machine lay in a revolving cylinder covered with metal spikes which tore at the cotton deposited on a grid, leaving the seeds behind and taking the cotton lint to the front. The machine was at first turned by hand, but was soon linked to water power. This vastly quicker and more efficient method of ginning soon put the United States at the head of producers of raw cotton.

For centuries in India and China, cotton was opened up and freed from dirt by bowing (5). The bow, a long pole with a gut string stretched from end to end, was suspended above the fibres from a springy stick attached to the wall. The bower, crouched on the floor, held the bow by a hand-grip with his left hand (protected in a piece of cloth on account of the friction). The pile of fibres to be bowed were on his right and as he hit the gut with a mallet held in his right hand he turned the bow away from him so that it flicked the fluffed-up cotton into a separate pile. After bowing, small portions of cotton were rolled between wooden bats to form thin rolls for spinning.

5 Eighteenth-century engraving from Diderot's *Encyclopédie* showing (top) a cotton bower; (bottom left) a detail of the hand grip and mallet; (bottom right) the cotton plant

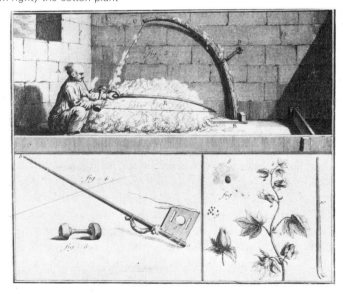

The Arabs became skilled in the manufacture of cotton, and when they penetrated into Spain in the 8th and 9th centuries it was one of the industries that accompanied them. By the thirteenth century cotton weaving was established in Christian Catalonia, and from Italy in the previous century there is a record of raw cotton being weighed on the public scales in Genoa. About that time cotton cloths are known to have been woven in Italian cities, for instance Bologna, and by the thirteenth century it was an established industry in Venice (*Ciba Review* no. 64), after which it spread northwards into Germany and Flanders yet not reaching England until the very beginning of the seventeenth century.

However, all cotton goods made in Europe prior to the Industrial Revolution were made with linen warps because of its strength, and the cotton used only in the weft. This produced a type of cloth known as fustian and is believed to have been made in a town with a similar-sounding name outside Cairo.

WOOL

On primitive sheep short, fine wool was found under a protective outer coat of coarse hair, but by selective breeding over the centuries the two extremes have been eliminated to produce wool with more uniformity. Wool fibres grow in groups forming locks known as staples; the length of these can vary from approximately 1in. to 18in. (2·5–45·5cm.). The main body of an individual fibre, the cortex, often hollow in the middle, consists of millions of long spindle-shaped cells. As well as giving strength these cells are linked together in such a way that when the fibre is stretched the links unfold and when released return to their original position. This formation is the primary factor which gives wool its elasticity and springiness. The cortex is encased by the scale-cell layer, a series of horny irregular cells which overlap from root to tip (like a tiled roof). It is these scales that give wool a tenacious property which facilitates spinning. Their organisation can vary considerably, but generally speaking there are more scales on fine wool than on coarse. The whole fibre is protected by an outer sheath, a membrane which is water-repellent but allows the fibre to absorb moisture through microscopic pores, and this generates heat. A second factor which gives wool its elastic property is its wavy structure known as crimp. This not only forms up and down a fibre but spirals round it. Crimp also helps to hold the fibres together in spinning, and at the same time prevents them from lying flat. Lustre is the shine of the fibres which reflects light and is found more on the long wools which have the larger scales, are coarser, and usually have less crimp.

During spinning the fibres undergo stretching, a little of which is sometimes retained. When the yarn is wetted the tendency of the fibres is to return to their original state, causing shrinkage. Also, felting and shrinkage occur when the fibres are compressed and relaxed in water, which causes them to become looped and entangled. This takes place more readily with fine wool with an abundance of scaliness and springiness particularly if the wool is softly spun which allows the fibres greater freedom of movement. It is less likely to happen with coarser wool which has been tightly spun.

As the fibres grow the animal exudes both grease and suint (sweat) which covers the fibres and is known as yolk. As a certain amount of lubrication is necessary for smooth spinning it is possible to spin wool in its natural state. The grease contains lanolin, which is pleasant on the hands, but some fleeces have an excess of yolk which forms into sticky yellow bobbles around the base of the staple and can be a hindrance in the spinning of unscoured wool.

Wool is removed from the living sheep with shears, following the contours of the animal in such a manner that it comes off in one piece (6). This is the fleece, sometimes called clipped wool, and it can vary in weight from $1\frac{1}{2}$ to 18lbs. (·680 to 8·154kg.). If a sheep has died, the wool is known as dead wool or fallen wool; it can be plucked off the animal. Wool also comes from slaughtered sheep and according to the method used to remove it is known as fell, skin, or slipe wool, and is usually of inferior quality.

6 Illustration for the month of June from a fifteen-century calendar. British Library manuscript, Add MS. 18855, f.109. Sheep-shearing in the Middle Ages.
Photograph: Reproduced by permission of the British Library Board

Hand shears (*figure* 1) had been in use since Roman times. After power became available shears were sometimes worked by a small steam engine and nowadays one sheep can be shorn with electric shears in $2\frac{1}{2}$ minutes. Shearing is done in early summer and should coincide with the moment when the animal is ready to moult. In the past it was done by both men and women; in June 1870 Queen Victoria described in her *Journal* sheep-shearing in Scotland done by women: the sheep first had their legs tied together by the shepherd who 'then placed them on the laps of the women who were seated on the ground, and who clipped them one after the other, wonderfully well, with huge scissors or clippers . . . and Mrs Morrison, who seemed rather new at it, and had some difficulty with these great heavy sheep, which kick a good deal. . . . The clippers must take them between their knees, and it is very hard work. . . . It was a very picturesque sight, and quite curious to see the splendid thick wool peel off like a regular coat.'

The fleece is then folded with the clipped ends outermost, the sides to middle, and rolled from the tail end into a bundle with the neck wool pulled out and twisted round to form a rope. This more than encircles the bundle so that the end can be tucked in. A well-rolled fleece can be carried by this rope of wool with no fear of it falling apart.

Wool usually needs to be sorted to separate the coarser wool from the fine and the longer wool from the short, matching them into grades that have the same characteristics. It is usual to find several qualities of wool on each fleece. In Britain the relative fineness of the fibre is known by its quality count. The higher the number the finer the wool. Today an arbitrary scale, it developed from the worsted system of the number of hanks of 560 yards of yarn that could be spun from 1lb. of wool. The range is from 28s to 100s, the latter only ever reached by the merino breed. (This scale should not be confused with yarn counts from which the system developed.)

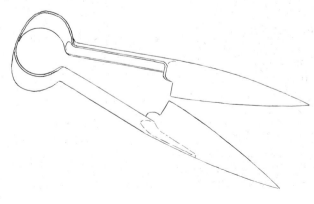

Figure 1
Sheep shears

The merino, thriving in dry lands and in large wandering flocks, produces a fine crimpy soft wool, often referred to as botany wool after Botany Bay in Australia. It is a breed that does not flourish in Britain because of the damp climate, although attempts were made to introduce it into Scotland between the fifteenth and seventeenth centuries and into England in the eighteenth century. British breeds fall into three main groups: those living in dry warm areas produce short fine wool, those in wet bleak areas, long lustrous wool, and those that exist in high exposed areas usually have rather coarse wool lacking in crimp and lustre. Some in the last group resemble primitive sheep and have an inner and an outer coat. (A list of some different breeds is given in the Appendix.)

Improved breeding, particularly by Robert Bakewell in the eighteenth century, makes a knowledge of medieval wool difficult. The words 'long' and 'fine' used to describe wool when fleeces weighed as little as $1\frac{1}{2}$lbs ($\cdot$680kg.) may not mean as long as we would think today, since, broadly speaking, we connect long with coarse and fine with short. Early in the Middle Ages, Flanders and Florence outshone all others for the making of fine woollens, but both these centres depended for their supply of wool on other countries, mainly England but also Spain and Burgundy. The Flemings were the first to buy wool from the English, but the Italians quickly followed suit and soon outstripped them as wool merchants. The extreme fastidiousness of the Flemings over wool sorting is considered one of the reasons for their success as cloth manufacturers; the mixing of different kinds of wool was strictly forbidden (*Ciba Review* no. 14). It is therefore rather surprising to find that in England this task was done by quite small children.

After sorting, the wool was scoured to remove dirt, and then rinsed in fresh water, tied in a bag or put in a basket or cage (**7 i**), wrung out and hung up to dry. In an article that appeared in the *Universal Magazine* in August 1749 about the wool industry, the following was said: 'To use your wool to the best advantage, you must scour the fleeces in a liquor, a little more than lukewarm, composed of three parts of fair water, and one of urine; and then, having continued long enough in the liquor to dissolve and loosen the grease, it is taken out, drained, and washed in running water. You will know when it is well scoured, when it feels dry to the touch, and has only a natural smell. In this state it is hung out to dry in the shade, the heat of the sun being apt to make it harsh and untractable. When dry, it is beat with rods on hurdles of wood to clear out the dust and grosser filth; the more it is thus beat and cleared, the more soft it becomes, and the better it spins; after beating it is well picked, to clear what has escaped the rods. In this state the wool is fit to be oiled with one fourth of its weight, if it be designed for the woof [weft]; but only one eighth, if it be for the warp.' The beating was sometimes called willowing (**7 iii**). Without the oiling, the natural grease made by the sheep having been removed in the scouring, it is not possible for the fibres to slip sufficiently to spin an even yarn. A variety of oils were used such as sheep tallow, melted pig fat, butter or goose grease.

7 Detail showing six pictures from a series of 16 depicting the processes of making a woollen cloth known as 'Laken'. Painting by an unknown Dutch artist c.1760. i. scouring wool, ii. dyeing, iii. willowing or beating with rods, iv. stock carding or carding bench, v. spinning, vi. winding bobbins or spools on a closed ended winder; in the background, a warping mill and creel. *Photograph: Central Museum, Utrecht*

From this stage onwards wool can be prepared in one of two ways, either with combs or with carders, which respectively gave the terms used to describe these two methods.

When combed – the earlier of the methods – the wool passes through the teeth of the comb which removes any short fibres and aligns the rest. When spun, these parallel fibres come to lie close to one another and flatter, so that with their ends caught into the twist of the main core of the yarn the result is smooth, cooler and often harder; this we know as worsted yarn.

When carded, on the other hand, the wool is only roughly aligned (no short fibres being removed) and it becomes mixed and criss-crossed, thus retaining maximum bounce. When spun, air is trapped between the fibres making a lofty insulated yarn which is very warm. With the longer fibres forming the main core of the yarn and the short ends not smoothed down, it also makes it hairy, and often softer; this we call woollen yarn.

In the *Statutes of the Drapers of Chauny* of 1410 it was stated: 'Every comber must pull only one third of combed wool leaving two thirds for weft'; other references of the time say it should be half and half. At this period combed wool was associated with warp yarn so we see it was the longest wool that was used. In a fifteenth-century illustration (8) a lady and her two servants are preparing and spinning wool. The woman on the right is combing wool, using a set of vertical spikes set in a wood block, not unlike a flax hackle. She has a staple in each hand and is drawing one of them across the spikes, deliberately keeping the fibres parallel. The results of her care can be seen quite clearly on the mistress's distaff from which she draws downwards from the bottom of the prepared staples to make the worsted-type yarn. When the wool is combed, aligning the long fibres, there are, of course, the shorter ones which get caught in the teeth of the comb; but wool was too valuable to waste, so these short fibres, known as noils, were removed and given to the carders to prepare for woollen spinning, probably for use as weft. The second servant in the picture sits with a pair of carders. In the manuscript of the Decretals of Gregory IX (British Library, Royal 10. E. iv) there is a miniature of a woman combing; she is sitting and using two combs, each with a single row of teeth, one of which is propped against a stand so that both hands can manipulate the second comb, which gives the impression that it was heavy. She may be doing it for domestic requirements only, since in the centres where the manufacture of cloth required worsted yarn and therefore combed wool, the work was done by men. It was considered a hard and unhealthy kind of job, and as the process of combing was one of the last to be mechanised there was still work for wool combers as late as the 1850s.

Wool combers were journeymen and seemed to have been a law unto themselves, since they might arrive on a Monday morning and yet not start work in earnest until later in the week. According to the *Book of Trades* of 1815 'The business of the wool-comber is different in different counties; some, as the wool-combers in Hertfordshire, prepare it only for worsted yarn &c; others, as those in and near Norwich prepare it for weaving into camblets and other light stuffs.' The fact that the 'business' varied from county to county could be the reason for contradictory accounts of wool-combing methods. 'He first washes the wool in a trough', continues the writer in the *Book of Trades*, 'and when very clean, puts one end on a fixed hook, and the other on a movable hook, which he turns round with a handle, till all the moisture be drained completely out. It is then thrown lightly out into a basket. . . . The wool-comber next throws it out very lightly into thin layers, on each of which he scatters a few drops of oil; it is then put together closely into a bin, which is placed under the bench on which he sits.' This confirms other descriptions that the wool was oiled while still damp.

The combs weigh about 7lbs. (3·17kg.) each (9); the teeth, made of highly

8 Fifteenth-century manuscript, Royal MS.20 CV, f.75. From right to left, wool combing, carders, and spinning with a spindle from combed wool attached to a distaff.
Photograph: Reproduced by permission of the British Library Board

9 A pair of wool combs.
Photograph: Castle Museum, York

tempered steel, are set into horn at an angle of around 80° to the handle. The number of rows of teeth vary from three to nine according to the type of wool being combed. The teeth of each row are progressively longer, increasing from about 7in. to 12in. (18–30·5cm.). The width of the comb is approximately 7in. (18cm.). (A blunt tooth would be sharpened with a polishing file and a bent one straightened with a small brass pipe.) The stout handle, usually made of ash about 1ft. (30·5cm.) in length, has two steel-lined holes, one at the side and one at the end.

The combs were warmed on a charcoal stove which was sometimes made of unbaked clay but often of metal, with a stone slab above leaving sufficient gap for the teeth of the combs to be inserted. Some stoves had, below the slab, a plate on which the combs rested. The stoves held four combs, for it was usual for four combers to work together (according to the *Book of Trades*), each with his separate gear, so there were four pairs of combs, four benches, four bins to hold wool and another four for the noils. The combs were heated so that they would slide through the oiled wool easily and not damage the fibres. The heated comb was first put on to a jenny, a wooden ledge attached to a post which held the comb with the teeth pointing upwards. The wool was lashed on by throwing small quantities at a time on to the teeth of the comb so that as much of the wool as possible formed a fringe in the front of the comb. When it was about half full, the comb was transferred to a pad (a metal bar) also attached to a post, which had two metal spikes corresponding with the holes in the handle, to hold the comb in a nearly horizontal position. The comber took the second comb, also warmed in the pot, and with both hands, swung it, rather like a chopper, into the fringes of the wool, each time getting deeper in and nearer to the stationary comb. This was known as jigging. When sufficiently jigged, the wool was pushed to the spring of the comb (about half way along) and, again using both hands, the comber drew it off into a continuous 'rope' called a sliver (10), which was then divided into 4ft. (122cm.) lengths known as fingers. These were laid overlapping side by side to mix the shorter fibres which came off last with the longer ones that came first. After the drawing-off, the fibres left at the front of the comb (milkings) and those at the back (backings) were added to the next lot of jigging (the noils being discarded and placed in their bin). The process was repeated until all the short fibres were removed. Sometimes the final sliver was pulled through a horn ring known as a diz which both smoothed the fibres and kept the sliver to a uniform size; this was then wound into a ball called a top. In the *Universal Magazine* article of 1749 the combers are described as sitting with the combs on their knee, the teeth vertically up, for lashing on. Seventeenth- and eighteenth-century engravings of wool combers in Germany and France also show the combers sitting, although standing to pull the slivers from a comb attached to a post.*

Wool combing was done in rural areas in places as far apart as Scotland and Greece, not as an organised trade but by women holding the combs one in each hand (rather like carders). These combs were of a much lighter construction than those used in the trade, with a single row of teeth set in horn and either straight or curving slightly towards the handle. One was used to comb the wool and the other to hold it; in Denmark they sometimes fixed the comb holding the wool on a bracket attached to a post. (Adofs-Utställningen 1933; H. P. Hansen 1947; I. F. Grant 1961). In Greece, however, one finds combs with two rows of teeth.

Carders are first recorded, in France, late in the thirteenth century. Carders were used in pairs, like combs, and may even have been developed from them. The name is presumed to be connected with the Latin *carduus*, a thistle, whence it is now often assumed that thistles (or teasles) had been previously used, though

---

*The patron saint of wool combers was Bishop Blaise (c.289–316) who had no connection with textiles but is said to have met his death by having his flesh torn with iron combs similar to those used by wool combers. In Bradford, Yorkshire, a festival took place every seven years to commemorate the saint. This started there in 1769 (there had already been one in Leeds in 1758) and continued until 1825. It was a colourful procession on horseback with an abundance of wool, some of it dyed, being carried. Riding along were wool staplers, worsted spinners, wool sorters, charcoal burners, dyers, comb makers, apprentices and wool combers wearing full-bottomed wigs made of combed wool. Bishop Blaise himself was represented surrounded by shepherds and shepherdesses, Jason and Medea and a company of guards. A smaller-scale affair was attempted in 1857.

10 After an engraving of a wool comber
drawing-off'. Book of Trades, 1815

it seems equally likely that carders received their name for their resemblance to thistles. Attempts to card with wild teasles, *dipsacus sylvestris*, prove unsatisfactory since the spikes stick to the wool and do not pass through it. If thistles or teasles had been used, it seems more feasible that it was simply for opening up, teasing, or picking wool (remembering that unless carefully handled, wool easily becomes somewhat matted after scouring) in preparation for carding or combing; even so, one feels that fingers do the job better.*

The carders (or cards) are made of two flat pieces of wood, each with a handle and varying in size from large and square (see **8** p. 33) to small and oblong. On one side of each of these carders are hundreds of small wire hooks set in leather, the hooks bent towards the handles though early illustrations (such as **8**) show the wires quite straight. Small tufts of wool are placed on one carder, which is held in one hand resting on the knee while the other carder is drawn across it several times, the hooks therefore going against one another. The wool is transferred from carder to carder and finally eased off the hooks. The fibres, all adhering together, come off as a sheet of wool which is rolled into a spongy cylinder, called a rolag. The fibres are thereby coiled and when they are drawn out by the spinner they go into a spiral so that the maximum air is trapped.

It should be noted that in the Middle Ages short wool was treated very similarly to cotton. In the fifteenth century wool workers in Constance, Germany, complained of the bow being used for cotton when it should be used for wool (Singer, Vol. II). Conversely, when cotton was prepared in England in the seventeenth and eighteenth centuries it appears that it was scoured (although this was not necessary), beaten and carded.

An improvement on the hand carders were the carding benches or stock cards (see **7** iv p. 31). The lower card slipped into slots on the raised sloping top – the stock – and the oiled wool was kept in the recess underneath. The person carding sat astride the bench to manipulate the second card by hand; alternatively, some benches were constructed so that the sloping stock was attached at the bottom to a stool which allowed the person to sit with knees together facing the fixed card (in Wales, for instance, and also Holland), a piece of carding cloth being nailed to the stock. Because the lower card was fixed it was necessary to change the grip on the hand card to transfer the wool from one card to the other, which need not be necessary with ordinary hand carding.

> 'The Stockcarder his arms doth hard employ
> (Remembering Friday is our market day)
> The Knee-carder doth (without controule)
> Quickly convert it to a lesser roule.'

*The fuller's teasle, *dipsacus fullonum*, has hooks on the ends of its spikes and is equally ineffectual for carding. When used for nap-raising several teasles, forming a thick surface of spikes, were fitted into a frame with a handle. Sometimes these can be seen in a museum today and should not be mistaken for carders. Teasles were later fitted into the cylindrical drums of a nap-raising machine called a gig-mill.

From this, part of R. Watt's poem *The Young Man's Looking Glass* (1641), we see that the stock carders were worked by men and that hand carders were still used afterwards, but in the eighteenth century heavier and more elaborate stock cards were also used; the lower card was fixed to the stock and the upper one, because of its weight and size, was suspended from the ceiling and worked by a cord, pulley and counter-weight system. It was said that this doubled a man's output. James Hargreaves, the inventor of the spinning jenny, made improvements to stock cards used for cotton (Aspin & Chapman, 1964).

In the Low Countries, a bench had attached at one end a large comb made of coarse metal teeth, over which was a triangular construction holding a similar coarse-toothed comb which could be swung back and forward. This was used for making flock.

In Greece, carding benches used for wool had a low stock for the fixed card which was mounted on a flat board resting on the ground. The person carding could work either sitting on the board or crouching.

SILK

Silk is the only thread that is produced by the animal itself, the species being *Bombyx mori*, the caterpillar for the silk moth. (A spider's web is too gossamer-like to be of use in textiles.) Silk worms eat leaves of the mulberry tree, *Morus alba* or *Morus nigra*, the white variety being the most usual. The worm eats for 35 to 42 days, at the end of which it loses its green colour and becomes transparent and creamy-coloured. It is then ready to spin its cocoon, whereupon it attaches itself to straw with a small secretion of thick viscous protein solution. This solidifies when in contact with air, forming the silk thread which is round and smooth. The solution exudes from either side of the worm's head through an exit tube called a spineret, and thus it produces a twin thread which is held together by a gummy substance called sericin. The worm waves its head in a figure-of-eight motion, depositing layer upon layer of silk thread inside the cocoon, which takes about three days to make. In silk production the chrysalis is then killed by stifling with steam or heat to prevent it from changing into a moth and forcing its way out of the cocoon as this would break the filament of silk.

To prepare silk for use it is subjected to two winding processes. The first, to unwind the cocoons, is known as reeling. The beautiful thread is so fine that several cocoons are unwound together. When done by hand the cocoons are placed in a basin of boiling water to soften the sericin. The ends are found by stirring with a brush or rod and are reeled as one, the sericin making them stick to each other. The thread passes through an eyelet and over pulleys so that it crosses itself, thus removing surplus water and ensuring cohesion, before being wound on to a rotating frame or reel and formed into a skein. It is skilled work to keep the thread an even thickness, since no two cocoons have an equal length of filament.

The second process of silk winding is to twist several reeled threads together to form a stronger yarn. This is known as silk throwing. The tightness of twist given to the silk as it is thrown can vary, and accordingly has different names. Thus a low twist yarn suitable for weft is called tram, while the very high twist of crepe yarn gives the fabric of that name its puckered surface.

Before mechanisation, silk throwing was done with the same tools that could have also been used for plying, or for that matter for spinning.

Silk waste comes from three sources:

(1) Floss or blaze is the beginning of the cocoon; it is soft and fine but of poor quality and lacking in strength.

(2) Frisson or strusa is the waste from the beginnings of reeling; it is collected from time to time, stretched to straighten the threads, and dried.

(3) Bassinet comes from broken and pierced cocoons, double cocoons, the stained cocoons of dead worms, and the last part of cocoons after reeling.

Silk waste was used by the Chinese as wadding for the lining of garments and bed clothes. After boiling the silk to de-gum, and opening up broken cocoons with a thumb nail, the fibres while still wet were 'put under small bamboo bows', which made them into a white floss. This was also used for spinning. 'The Hu-chou silk woven from yarn spun from this silk floss fetches a rather high price' (Sung Ying-Hsing, seventeenth century).

The making of silks was another skill (besides cotton manufacture) which the Arabs took with them to Spain in the 8th and 9th centuries. About the same time they also took sericulture to Sicily, after which Roger II, when he was reigning King of Sicily in about 1130, established silk manufacture at Palermo and also in Calabria in the toe of Italy. In the second half of the thirteenth century pierced cocoons from Calabria were being spun in Lucca (*Ciba Review* no. 80) where there was a flourishing silk industry as there was also in Florence and Venice.

Chapter Two

Spindle wheels

Before embarking on the spinning wheels themselves it might be as well to take a brief look at the first tool that was used for spinning, namely the spindle (*figure* 2). If in the first place this was nothing but a stick on to which the yarn was wound, at some period in history a whorl (sometimes called a whirl or wharve), a disc-shaped weight, was fitted on to the stick to act as a fly wheel. The word 'spin' stems from a word which means to draw-out, while 'whorl' means to whirl round.

The spindle was made in all shapes and sizes, but basically it is a straight stick made of wood or bone, often with a hook or notch at the uppermost end. The whorl, usually made of wood, clay or stone, can be fitted on to any part of the stick, although it is more often found near to one or other of the ends. On occasions it is shaped as part of the stick. When starting to spin with an empty spindle it is necessary that a length of spun thread be tied to the stick.

To produce the twist needed to make a thread from fibres using a spindle, one of the more familiar methods is to start its rotation by giving the stick a sharp turn with one hand (or by rolling it along the hip). It is then suspended in the air by a short length of spun yarn hitched firmly round the stick, notch or hook, the whorl keeping the spindle vertical and giving it momentum. Meanwhile, the other hand holds the yarn firmly at the point where it meets the fibres, so that the twist will not immediately run into them. Once the spindle is in motion the spinner draws-out, or drafts, a few fibres at a time from the prepared mass held either in the hand or on a distaff (see plate 8). The drawn-out fibres are attenuated between the hands and the twist to form them into thread is allowed to slip through the fingers of the hand nearest the spindle. The repetition of drawing-out and twisting continues until the spindle stops rotating (and before it reverses its direction) or almost reaches the ground. It is then picked up, and the yarn unhitched and wound round the stick close to the whorl, one hand guiding the yarn and the other turning the spindle.

The advantage of the spindle is that once it has been set in motion the spinner has both hands free to draw-out the fibres and control the twist, which, when it reaches the fibres, distributes itself along the full length of those drawn-out. If this is done unevenly the thinner places will be more tightly twisted (i.e. the angle of the twist will be more acute) and hard, while the thicker places will be under-twisted and soft (commonly known as slubs). The skill of the spinner lies in drawing-out the fibres evenly by maintaining the correct balance between the speed of draw, the quantity of fibre and the amount of twist.

The weight of the spindle varies with the length of fibre, the thickness of the thread required, and the amount of twist it is given, since a fine, lightly twisted thread cannot support a heavy spindle, but a thick thread can bear more weight which will give extra momentum. In India, cotton can be spun so fine that the tip of the spindle may rest on the ground or in a coconut shell so that it is constantly supported as it rotates and there is no weight that might break the thread. In hot climates the spinners keep bowls of chalk beside them in which they dip their fingers to prevent the fibres sticking to them.

The spindle, this simple device, can be used to spin every sort of fibre and to such perfection that it is wonder that it was ever relinquished at all in favour of the spinning wheel. It was cheap to make, easy to store, and handy to carry around so that spinning could continue in conjunction with other work, indoors or out.

Of the many methods for using a spindle there are two which are perhaps worth noticing with regard to the action of the spinning wheel. One, practised by the Navajo Indians for example, uses a spindle sufficiently long for one end to rest on the ground when the spinner is in a sitting position; the uppermost end is without a notch and has a fairly pointed tip to which the yarn already on the spindle spirals. The spindle is kept in motion by rolling it along the hip in one

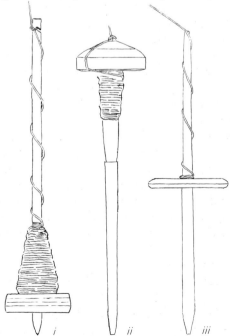

Figure 2
Spindles. (Examples after:
i. contemporary English ii. Cyprus iii. Sudan)

direction and the twist is given by the formed yarn slipping over and off the tip of the rotating spindle. The other method was seen by Frödin-Nordenskjöld in use by the women of the Chimane tribe in the Andes of Bolivia. The spindle was supported at one end between the toes and the other end rolled with the hand on a block of wood. Thus the spindle was placed horizontally to the ground (A. Linder 1967).

The first type of spinning wheel is unquestionably the spindle wheel. This came to have many names, but when referring to them collectively here the term spindle wheel will be used (though others have called it the hand-turned spinning wheel) to distinguish it from later types of spinning wheel. It is sometimes called the one-thread wheel, possibly to avoid confusion with the multiple spinning appliances of which it is the precursor, or possibly because of its single driving band. Because it was used in the main for cotton and the shorter stapled wools it is occasionally referred to as the short fibre wheel.

The components of a spindle wheel are mounted on a bench (*figure* 3). At one end of this, one or two uprights support the horizontal spindle with two bearings. The whorl, made of wood or metal and now no longer necessary as a weight, is firmly attached to the spindle and lies between the bearings. The circumference of the whorl is grooved to take the single band which links it to the driving wheel positioned at the other end of the bench. The wheel is either supported between two uprights or attached to a single post. To make it possible to move the spindle further from the wheel to tighten the driving band, some method of adjusting the tension (known simply as the tensioner) is often included in the design. If the band is too slack the spindle will not turn.

In general the spinner places the spindle wheel along her right-hand side with the spindle in front and the wheel beside her. She turns the wheel with her right hand while in her left she holds either a rolag or a bunch of prepared fibres. As she moves her hand away from the spindle tip she releases part of the rolag, or

Figure 3 *Spindle wheel (after an example from the Walloon parts of Belgium). Spindle with a disc at the base; bearings made of straw; whorl with several grooves; no tensioner*

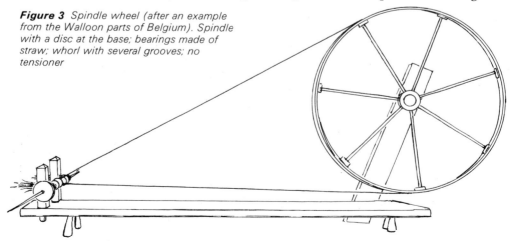

some of the fibres, which become attenuated under tension and formed into yarn by the twist given by the rotating spindle and the yarn slipping off the tip (as with the Navajo spindle). It is usual for the fibres to be drawn-out at an obtuse angle to the spindle; an angle nearer in line with it would give a looser twist and an angle nearer to 90° would make it tighter.

When the spinner has drawn-out to the capacity of her arm span the formed yarn may be held, while the wheel continues to be turned, to give more twist for strength. It is up to the spinner to judge how much twist is required, according to the purpose of the yarn. Once the thread is made, the spinner reverses the wheel a little, at the same time guiding the yarn off the spindle tip; this is known as backing-off. Then, turning the wheel in its original direction, the spun yarn is held at right-angles to the spindle and guided to its base, where it is wound on and built into either a cone shape or a cop (*figure* 4).

Yarns can be spun in either direction. The yarn spiralled to the tip of the spindle must be in the same direction as its rotation. When the wheel is turned to the right (i.e. clockwise) the spindle also turns to the right, causing the yarn to

Figure 4
i. Cone; Z twist; open band
ii. Cop; S twist; closed band

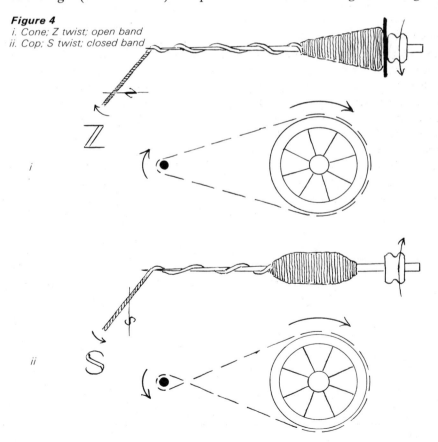

twist into a spiral from right to left towards you which is called Z twist. On the spindle wheel this is called open band. To turn the spindle in the opposite direction (anticlockwise) the wheel is not reversed, but instead the band is crossed over itself before it encircles the spindle whorl and this is called closed band. The spiral sloping from left to right towards you produces an S twist yarn. (With the flyer spinning wheel, on the other hand, the S twist, when required, is produced by turning the wheel anticlockwise, not by altering the band.)*

The spindle wheel, with its relatively large driving wheel and small spindle whorl, and hence fast rotation of the latter (even if the driving wheel is turned quite slowly), provides a quicker method of producing yarn than the spindle. Even so there are two disadvantages: (1) that spinning and winding-on are two separate processes; (2) that since one hand is engaged in turning the wheel, only one hand is available to control both the fibres and the twist. The spindle wheel had two other functions in textile technology which we have already noted: it could be used for twisting together continuous threads, either silk throwing or plying, and it could be used for winding yarn on to bobbins. To the historian of mechanical invention the spindle wheel is of interest on account of its belt transmission of power, but this is necessary for winding and twisting as well as spinning. The function of a spindle wheel is therefore often not discernible unless it is seen in use. A guide can sometimes be the length of the spindle, which if rather short probably signifies a winder. When used in this last capacity its disadvantages are of little consequence, since the hand not turning the wheel merely has to keep a tension on the yarn and guide it on to the bobbin. Throwing or plying needs more skill since, as in spinning, the tip of the spindle gives the twist, so the yarns must be fed to it at a uniform tension and be guided off for winding on; but this is not difficult to do with one hand. Spinning itself needs the most skill because of the double role of the one hand. One might assume that the easiest usage came first. Even so we are uncertain not only whether as a spinning wheel it was developed from the winder or from the spindle (it being a combination of the two), but also with which of the fibres it was first used. For these reasons its true origins may never be known.

SPINDLE WHEELS IN THE EAST

India is usually considered the birthplace of the spinning wheel as such, for the production of cotton yarn. However, Irfan Habib of Aligarh Muslim University states; 'whereas Sanskrit appears to lack any word for it, the term in use in most North Indian languages and the one also carried to Nepal is the Persian name for it "Charka"'. He also says that in texts where the spinning wheel might well have been mentioned, it is not, nor have any early illustrations come to light. On the other hand Joseph Needham in *Science and Civilization in China*

---

*Some people argue that if the thread is looked at from the other end, the direction of twist will change, but this is a fallacy: while clockwise and anticlockwise and left and right do change this is not so with S and Z; whichever way you look at the thread the direction of the twist is constant.

claims that the oldest representation of the spinning wheel yet known from any culture is a painting attributed to Chien Hsüan datable to about 1270. The picture shows a son saying farewell to his mother, who is sitting at a spindle wheel, turning the wheel with her right hand as she draws-out the fibres with her left. Undoubtedly the Chinese had highly sophisticated apparatus for their silk manufacture and they used as well the fibres from hemp, ramie (known as China grass, a type of stingless nettle) and cotton. It is possible, as Needham suggests, that the earliest silk, when obtained from the wild silk worm, needed to be spun, not reeled, and for this purpose the Chinese may have made the discovery of driving a spindle with a wheel. W. Born (*Ciba Review* no. 28, *The Spinning Wheel*) dates the invention to between 500 and 1000 A.D., and although from the Chinese picture dated 1270 one can suppose that the spinning wheel had been in use for any number of years before that, yet it does narrow the gap between its possible evolution in the East and its introduction into Europe, if pictorial evidence can be a guide. It seems strange that the spindle wheel should have stayed away from the West for so long, when there was no lack of contact between East and West through trading, and particularly so if, as Irfan Habib states, by 1257 the spindle wheel was in use in Persia.

Documentary evidence from these early periods is sparse owing to various possible factors. When spinning with the aid of a wheel was first evolved it may have been kept a secret within the industry concerned, in line with several other methods and tools of textile technology. There may also have been difficulty at first in persuading the people to use it. Then, once established, it took its place in everyday life, where it did not offer a subject for writers and artists. Even historians of textiles have often neglected to describe in any detail the techniques of spinning, or have over-simplified them.*

There are certain similarities between Indian and Chinese spindle wheels. In each case the wooden base set on the ground supports the spindle in bearings between two uprights at one end, and at the other end two uprights support the wheel by its axle. The wheel is turned by a crank handle on the axle (the spinner uses her right hand). The Indian Charka has two eight-armed stars of four shaped pieces of teak forming 'spokes' and mounted on the axle some distance apart (11). As there is no rim, the outer ends of these pieces of wood are interlaced with thread, thus forming a support over which the driving band runs, linked to the whorl of the spindle. Rimless wheels are still made in India and Pakistan; some have two discs of wood instead of the crossed teak pieces, with notches cut slantwise on the edges at intervals to hold the criss-crossed string forming the rim.

---

*J. M. Roland de la Platière, writing in 1780, tells us that spinning offers little for discussion. The movements are not very numerous, uncomplicated and a continual repetition, and it is in this exact repetition which is the fruit of the great skill. But he also says that as it is on this perfection that the beauty and quality of the cloths absolutely depend; as far as workmanship is concerned, one feels there is no operation more important.

11 Lucknow painting c.1830. An Indian
Charka being used for cotton spinning.
*Photograph: Reproduced by permission of the
Director of the India Office Library and Records*

The Chinese, on the other hand, have two sets of fine wood or bamboo spokes set into the axle, and the string forming the rim is tied or wound round the tops of the spokes. Varieties of these rimless spindle wheels can be found all over the Eastern hemisphere; in Burma, for instance, the spokes carry metal rings to jingle as the wheel turns. Indonesia has a wheel similar to the Charka, but the spinner holds the thread close to the spindle between her toes while she draws-out the fibres with her hand. None of these wheels have legs; the Indian Charkas are placed on the ground beside the spinner who sits either on the ground or on a very low stool, a natural position in a hot climate though not suitable to the damp and cold of northern Europe. Pictures of Chinese spindle wheels usually show the spinner sitting on a stool.

Needham also tells us that Chinese multiple-spindle spinning wheels were used with a treadle by the fourteenth century. The type of treadle is, however, of a completely different conception from the European one, and is therefore unlikely to have had any bearing on the adoption of the treadle here (12). The driving wheel differs from the Chinese type just described. A single upright supports the

12 Illustration from Nung-cheng ch'üan-shu, 1562—1633 (Ser.d. 107/15 f.13a & b). Chinese multi-spindle spindle wheel with universal joint type of treadle being used for spinning from balls of rovings.
Photograph: Curators of the Bodleian Library, Oxford

hub from which radiate a number of spokes each ending in a T-shaped block over the ends of which are attached two narrow hoops. The single driving band runs over the blocks and also over five spindle whorls which are set in an arc above the wheel. The treadle itself is a long pole, pivoted freely near one end on an upright post. The other end is tapered off and engages in a recess in one of the spokes. The spinner sits on a stool and puts both feet on the pole, one on either side of the pivot post, and pushing first with one foot and then the other, in the same manner as in riding a bicycle, the wheel is set in motion. In her right hand she holds a stick which is placed across the previously prepared ribbons of fibres, the rovings, as they are drawn-out. A picture contemporary with plate 12 shows that when three spindles were used instead of five, the stick was dispensed with and the fibres were drawn-out between the hands from short thick rovings which were wound round the left wrist. A photograph in the Science Museum, London, shows a woman using three spindles and the stick technique, which may indicate that long bast fibres were being spun. Spinning from rovings round the wrist may have been the method used for short fibres. The Chinese of course also used single spindle wheels, and the spindle and distaff.

THE ORIENTAL OR EASTERN TYPE
OF SPINDLE WHEEL IN EUROPE

Having in mind that spinning wheels are known to have been made in the same style from one generation to the next, to what extent can spindle wheels still seen in Europe today, or preserved in museums, reflect their early history? Firstly, wheels on which the driving band is not borne directly by a solid rim, in other words the rimless wheels already described, also came to Europe. It is however the wheel with a true hoop of thin wood (like the frame of a sieve) which is far more familiar in the West today.

In Greece (a link between East and West) rimless wheels driving small spindles with a swift beside them can still occasionally be seen in use as winders. The north, however, is still the textile area of the country and in the Ethnological Museum of Macedonia in Thessalonika there is a rimless wheel (the ends of the 20 spokes of each section joined by a circle of flat metal with a criss-cross of string between them) which drives a wood spindle (20in., 51cm.) incorporating several alternative grooves for the driving band. From the length of its spindle one would suspect that it had been used for spinning rather than winding. Endrie (*L'Évolution des techniques du filage et du tissage*, 1968) portrays a woman working with a rimless wheel in Bulgaria and the same has been seen in Yugoslavia. No doubt many more instances can be found in other East European countries.

Horner (1920) when he visited Italy remarked that there were no spinning wheels south of Rome. He must, however, have been thinking of the later types, since Paul Scheuermeier (*Bauernwerk* Vol. II, 1956) found spindle wheels far further south, although in Calabria they were only used as bobbin winders.

Spindle wheels were then made in Italy with both forms of rimless wheels as well as hoop rims and were used for winding, twisting or spinning (the last the most usual). In the province of Marche, according to Scheuermeier, there were a number of spindle wheels almost identical to the Charka, with criss-cross string to form the rim, turned by a handle and used for wool spinning. In both Tuscany and the Abruzzi there were spindle wheels reminiscent of the Chinese type with two narrow bands of wood or bamboo round the circumference. The wheels were quite large, in diameter approximately 30–36in. (76–91cm.), and mounted on four-legged benches. One example from Tuscany was turned by pushing the wheel round with the hand and was used for wool, but another example, in the Museo Nazionale delle Arti e delle Tradizioni Popolari in Rome (illustrated by Paulo Toschi, 1959) was turned by a crank handle; the wheel uprights extended below the bench to form legs while the legs under the spindle were shorter so that the bench sloped. A spindle wheel from the Abruzzi (Amatrice) had a large wheel with a hoop rim, and a spindle held in leather bearings. The structure rested on the ground and was being used for spinning by an old woman sitting on a chair. A few spindle wheels were found in Venezia in the foothills of the Alps but the type of rim is not specified, since Scheuemeier in his otherwise excellent distribution map does not differentiate between the hoop rim and the type with the two narrow hoops; but an example of the rimless Charka type was found in Forni di Sotto (Friuli) close to the Austrian border, and was again used for spinning wool.

In sixteenth-century Italy, Tintoretto's 'Harvest of Manna' shows a woman with a silk twisting wheel (illustrated in *Ciba Review* no. 29) beside her. This has a six-spoked rimless wheel turned by a crank handle and driving a spindle.

On entering Switzerland, whether through the Brenner Pass and Austria or over the St Gotthard Pass, one arrives in the area of speakers of the Romanisch (Rhaeto-Romance) language, with Chur the centre of the region. Here again can be found rimless spindle wheels with the zig-zag string joining the tips of the spokes. Plate 13 shows one from the Valley of Rheinwald, Graubünden, south-west of Chur off the road from Bellinzona, but other examples come from the area quite close to Chur, east and north in the foothills of the Alps over towards Austria. A photograph of an old woman from that area using one of these spindle wheels in 1930 shows her sitting, and using it for wool spinning. Schwarz (*Ciba Review* no. 59) states that the rimless wheels came to Switzerland with cotton in the seventeenth century. In the next century the Society of Cotton Spinners of Prättigua used this type of wheel (A. Linder, 1967), which could account for the number of wheels found in that area, even if in more recent times they were used for wool. The simplicity of its structure meant that this rimless wheel could easily be made at home by peasants. It is locally called the 'goat wheel' (*Spinnbock* or *Bockrad*) or Mother-of-God wheel. It is also said there that, because of its antique form, it 'dates back to the Bible' or as we would say 'came out of the Ark' (*Dicziunari Rumantsch Grischun*).

In Germany, Wilhelm Bomann, in 1927, illustrates a rimless wheel with a crank handle, then in the Celle Museum. He considers it to represent a very ancient form of spinning wheel although later made and used as a bobbin winder. A fine example of an Oriental-type spindle wheel in Germany is in the Deutsches Museum in Munich. It is known to be eighteenth century and comes from the region of Augsburg. The wheel is made with eight turned spokes and the rim of two narrow hoops joined with a zig-zag of ribbon $\frac{3}{8}$in. (1cm.) wide. The single post which holds the spindle can be tensioned by sliding it along a slot in the bench and tightening with a nut underneath. The whole construction is large, the wheel diameter $39\frac{1}{2}$in. (100cm.) and the bench approximately 55in. (140cm.) long, raised off the ground by four legs. The wheel is supported at the back by a single upright and is turned by a handle fixed to a spoke, not by a crank handle on the axle.

Such wheels also reached Scandinavia; examples with two narrow hoops over blocks of wood at the end of the spokes exist from near Stockholm and from further south in Ostergotland and Skåne. This wheel is there called the high wheel and is usually used standing, though the spinner sits when tired. Sweden was the only Scandinavian country which had a woollen textile industry of any importance; the earliest workshop was established in the middle of the sixteenth century and the craftsmen were probably German (M. Hoffmann, *The Great Wheel in Scandinavian Countries*, 1963). Norway and Denmark relied on domestic weaving and imports, but at the beginning of the seventeenth century the King of Denmark started a textile workshop in Copenhagen which was a cross between a house of correction, an orphanage, and a workshop for poor children. Dr Hoffmann tells us that the type of spindle wheel delivered to this establishment was known in Denmark as the Scottish wheel. It was confined to textile workshops and was never used by women of the upper classes or domestically at all. A *Scotroktos* or Scottish-wheel-woman was a woman of vice, and *skott* spinning means on a spindle wheel. As they were rimless, it seems more likely that they were introduced from Sweden than from Scotland, even though early in the seventeenth century there were Scottish textile merchants in that country.*

Hoffmann further says that in the middle of the eighteenth century the spindle wheel reached Norway but that in spite of the fact that it was used in a house of correction in Oslo, it did not get the same reputation as in Denmark, although it was made in imitation of the Danish wheel. It was used to a limited extent in

*John Howard, when he toured European prisons in 1777, noted that in Denmark all the wool spun in the several houses of correction in the Danish dominions was taken to the King's military cloth manufactory, built in 1760. 'In the Almindelige or great hospital there are nearly a thousand poor, mostly spinning worsted', to which he added a footnote, 'It was a hardship on the aged and infirm to be obliged to spin wool, when they had been for long accustomed to spin flax or hemp which is cleaner.' Howard also noted that in the hospital 'There was a room or two which belonged to a manufacturer (a Scotch gentleman) whose office it was, to give out and take in their work, and to pay for it'.

13 A European rimless spindle wheel from
Hinterrhein, Canton Graubünden, Switzerland.
Wheel diameter approximately 28in. (71cm.).
Photograph: Dicziunari Rumantsch Grischun

country areas where farmers had maids spinning on it, but it never came into
wide use. The spindle wheel preserved in the Nordsk Museum in Oslo came from
a house of correction. It has a wheel diameter of $35\frac{3}{4}$in. (90cm.) and an overall
height of 53in. (134cm.). The wheel, with two narrow hoops, has a criss-crossed
band of a felted woollen material with selvedges, which means it was made as a
strip. The spindle is made of a single piece of wood; of this $5\frac{1}{2}$in. (14cm.) is tapered
to form the actual spindle while the rest fits into the bearings, with a groove
between them to take the driving band. Its diameter is very small, little over an
inch (2·5cm.), which would make for a very fast rotation. It was used for short
stapled wool.

14 Interior from Näs Lerrgärd by Pehr Hilleström c.1775. Cotton spinning. *Photograph: Nationalmuseet, Dansk Folkemuseum, Brede*

At the end of the eighteenth century cotton spinning became fashionable in Sweden, and also in Norway where it was called English spinning. Pehr Hilleström, who portrayed everyday life in eighteenth-century Sweden, shows elegant ladies sitting to spin cotton on a spindle wheel with a criss-cross band forming the rim (**14**).

HOOP RIM SPINDLE WHEELS

It must not, however, be thought that these countries had rimless Oriental-type wheels only. Pehr Hilleström also painted women in Sweden using spindle wheels with the hoop rim, and in Germany too they appear in engravings and pictures. Moreover, on the more westerly side of Europe, and even looking as far back as the thirteenth century, there seems to be little sign of rimless wheels. There was of course no lack of contact, with trade routes from Italy across the Alps into Germany, Austria, Switzerland or through France going north into Flanders and from there by sea routes to England and further north. At the great annual trade fairs of Champagne in France and Frankfurt-am-Main in Germany the textile merchants sold their goods and there was opportunity for interchange. Venice, particularly, traded with the Far East through the Mediterranean ports and the Near East, and was a city where merchants of the East could meet those of the West.

An early written record mentioning spinning on the wheel in Europe is from Speyer, not far south of Frankfurt and on the route from Venice, and this forbade wheel-spun yarn being used for warp. To anyone familiar with spinning on the spindle wheel this is not surprising, since a softly twisted thread forms easily, but it takes a lot of extra twist to get it strong, particularly with very short fibres (which also takes more time) – a fact which may not have been fully appreciated by those early spinners. The need for such a regulation surely indicates that spinning on the wheel was an established method by that time (1298), though it does not give us any clue to the type of wheel or to which fibre was involved.

It has been suggested that the earliest type of spinning wheel first made its appearance in Europe with the Moors in Spain, although there seems to be little spindle wheel tradition there; indeed there seems to be some confusion in the terminology used, since *rueca* meaning distaff (rock) can apparently be mis-used as the word for spinning wheel (*torno* or *rueda de hilar*). In the South American countries so influenced by Spanish culture from the sixteenth century it is spindle spinning that is carried on as also in North Africa to which the Moors returned.

In Endrie's view it was with the Christian Catalan fustian weavers that the early spinning wheel (the spindle wheel) reached the fairs of Champagne even before the thirteenth century. Schwarz, and also Endrie mention a reference to cotton spinning by the hosiers (*bonnetiers*) of Paris in the middle of the thirteenth century but it is by no means certain that here the word *touret* signifies a wheel. If the spindle wheel reached France with cotton, then one would expect the wheel to have been of a rimless type, yet depicted in the windows of two northern French cathedrals built in the thirteenth century, Amiens and Chartres, are wheels beside looms being used for winding bobbins, and these have hoop rims. Among the donors of money for these French cathedral windows were weavers, who in this non-silk, non-cotton manufacturing area, were making woollen goods.

Since spindle wheels of the Far East do not appear to have the hoop rim, could this have been introduced somewhere on the way westward, or even in Europe itself? To reach France the spindle wheel, in some form, could have travelled northwards from Italy where, as we have seen, they not only had a spindle wheel tradition but also early established cotton and silk industries. Even if *en route* its use had been for silk winding and cotton spinning, it would seem that it was in the North, France or Flanders, that it was discovered to be suitable for spinning short stapled wool and for this purpose was made with a hoop rim.

In the British Isles, Flemish textile workers, encouraged to settle early in the fourteenth century, brought with them improved implements, one of which may have been the spinning wheel. That these were the people who introduced it seems probable, particularly when we are told that one of the folk names for the spindle wheel in the Isle of Man is the Flanders wheel (I. M. Killip, 1965).

The first known pictorial records of spinning wheels in Europe appear early in the fourteenth century in an illuminated manuscript of the Decretals of Gregory IX (thus only about 40 to 50 years after the Chinese picture mentioned earlier in this chapter). Written in Italy, probably for use in France, it was illustrated in England. Its delightful miniatures form graphically told stories, the six depicting spinning wheels coming in a group of 23 which also includes the one already mentioned of a woman combing (p. 32, chapter I), and others of her carding. While the woman is spinning, the pictures show the man doing various household duties, but towards the end he does the carding and finally has a turn at spinning, for which, it appears, he is soundly scolded. The spinning wheel is each time drawn more or less the same, and shows a medium-sized wheel (it could be estimated about 24in., 61cm. in diameter) with a hoop rim and supported between two wheel uprights (15).

A second manuscript with an illustration of a spindle wheel is the Psalter commissioned by Luttrell of Irnham in Lincolnshire, written and illustrated in East Anglia between 1335 and 1340. This miniature shows a larger type of driving wheel than the Decretals manuscript; in fact on its short little legs it stands higher than the spinner. It too is drawn to indicate a hoop rim but shows a plyed driving band crossed (only just discernible) to give an S twist to the yarn (16).

These pictures establish that from their earliest known time in Britain spindle wheels were made with hoop rims, a tradition that was rarely broken. Even though their size varies, there can be no doubt that the spindle wheels in both these manuscripts are being used for wool spinning, not only from the presence of the carders but also from the immensely long draw-out from a rolag, which clearly fascinates both illustrators. Neither of these wheels has a crank handle; instead, the spinner pushes it round either with her hand or with a small stick, later known as a wheel boy, wheel dolly, or wheel finger. In both illustrations the woman stands to work thereby gaining more space for drawing-out the fibres.

15 Early fourteenth-century miniature from Royal MS. 10 E IV, f.146. Woman spinning and man carrying loaves.
Photograph: Reproduced by permission of the British Library Board

16 Miniature from Luttrell Psalter 1338. Add, MS. 42130, f.193. Spinning and carding.
Photograph: Reproduced by permission of the British Library Board

From subsequent pictures and engravings we can see that the large hoop-rimmed spindle wheels, measuring at least a yard or metre in diameter, were widely used in the woollen industry in Europe, their benches resting on the floor at the spindle end to give maximum capacity for drawing-out, and the wheel being pushed round by the hand, though it must have been tiring for the spinner (17). We gather from the caption to plate 17 that in the eighteenth century there was no longer prejudice against this wheel for making warp yarn but that the method of carding was all-important. One could guess that the rolags were made to such an exact size that the spinner could use a complete rolag each time. By bending to reach nearly to the tip of the spindle, but not so close as to make joining difficult, she could draw-out as far as possible yet not lose touch with the wheel. A carefully and correctly judged build-up of twist or some other facet of technique, learnt from constant practice and experience, would have made it possible – but this is purely speculation.

Examples of spindle wheels preserved today show variations in both size and construction. In France, for instance, where the wheels are usually quite large, though always with the hoop rim, some have features pointing to a style reminiscent of the Oriental wheels. A noticeable characteristic is that the driving wheel is invariably placed between two wheel uprights, while in other parts of Europe this form of support was dropped in favour of a single post. In some areas as Lozère, Aveyron, and the Auvergne towards the south, and Maine-et-Loire in the west, the wheels are turned by a crank handle on the wheel axle, often from a sitting position, despite their large size, while in the north, Normandy and Brittany, the wheels are pushed round with the hand.

On the other hand spindle wheels often display similarities which are widespread; the use of straw bearings to hold the spindle was quite common in countries as far apart as India and Scotland. It was thought that the spindle ran more freely in straw since its silicious surface would give least friction. An example from Normandy has the straw doubled over the spindle and tightly bound behind the spindle upright, whereas in Lozère an example shows the straw plaited and then doubled through the upright. Another method, also in Lozère, uses bearings made from multi-stranded plaited straw forming two tabs approximately 3in. (7·6cm.) wide and slotted through the uprights; a small peg at the back of the upright holds each in place while the spindle is pushed through the tabs. Where straw bearings are used there is seldom a tensioner, but with such easily manoeuvrable bearings the driving band can be put on with the plaits slack and then tightened for tension; or in the case of the tabs, the spindle itself can be moved simply by pushing it through them further from the wheel. A spindle wheel from the Walloon part of Belgium (see *figure* 3) has the spindle held in place with wrappings of straw, and with its short legs and medium sized wheel is reminiscent of those illustrated in the fourteenth century.

Leather too was used for bearings and could also act as a form of tensioner since the loops of leather, passed through the upright, were held in place with a pin which could be moved to various holes punched in the leather, thus altering the position of the spindle.

Various other means of tensioning can be found, some very simple and some sophisticated. If the upright holding the spindle is a single plank, wedges can be pushed into the slot which holds it, to one side or the other in order to tilt the upright towards or away from the wheel. Sometimes the uprights bearing the spindle are mounted on a wooden block which can slide along the bench and be tightened with a nut underneath (as noted already in Germany, p. 50, and in plate 13).

Other spindle wheels have an arrangement by which a sliding batten is slotted into uprights which tilt away from the wheel; this also means the height of the spindle is altered. A more elaborate system has a second upright beside the spindle upright with an endless screw joining them and provided with a handle. Sometimes the handle points away from the wheel and sometimes towards it, it being easier for the spinner to reach it in the latter position. Another form of tensioner is fitted inside a small 'box' built round the base of the spindle upright. A wooden screw, parallel with the length of the bench, runs through a slot at the bottom of the upright and is fixed to the ends of the box. Two nuts, one on either side of the upright, can alter its position; sometimes there is only one nut,

17 From a copper engraving, 1728. The spinning room in the Waldstein Woollen Mill, in Oberlentensdorf, northern Bohemia.
1. The wool is weighed by the spinning master, turned over to the spinsters and yarn taken in return. 2. The wool for weft yarn is carded on the flat cards. 3. On the knee cards it is carded for the warp yarn for the spinners.
4. Wool is spun on large wheels. 5. Store of yarn.
Photograph: Deutsches Museum, Munich

so that the upright is held by the pull of the driving band.

The length of the spindle itself varies but 5–10in. (13–25·5cm.) is usual. It tapers to quite a narrow point and is generally made of mild steel though sometimes of hardwood. On some older wheels with wooden spindles there is a spiral groove either caused through wear by the yarn as it was spiralled to the tip, or else, equally possible, cut in order to guide the yarn there. Schwarz found in Normandy hand spindles cut in this manner for plying, though he could have mistaken worn grooves for cut ones. Metal too can become grooved. Discs at the bottom of the spindle against which the yarn is built into a cone are found on spindle wheels of many countries including India, Scotland, Sweden and Belgium.

The whorl, particularly a wooden one, can extend along the spindle for nearly the whole distance between the two bearings (*figure* 3). Such whorls are often conical in shape and can have as many as five grooves, sometimes all of the same diameter and sometimes differing. In the former case they are most likely to have been used for moving the band from one groove to another as a groove became worn; the spinner, working long hours, was thus not obliged constantly to obtain a new whorl from the local turner. With the width of the hooped-rim wheels there is no problem over alignment of the driving band to these different positions of the grooves.

To make a hoop rim two strips of wood, $\frac{1}{8}$in. or $\frac{3}{16}$in. thick (3–4mm.), 2in. (5cm.) wide on average, each measuring half the circumference of the finished wheel and allowing from $2\frac{1}{2}$–3in. (6·5–7·6cm.) extra for overlapping, are cut so that the straight grain goes round the wheel, square at the end edges and bevelled on the sides. The strips are soaked in running water (the time of soaking seems to vary from two hours to two weeks according to the maker and the type of wood). They are then steamed and bent into shape and placed round a mould with a rope tourniquet round the outside and with extra pieces of wood to cover the joins and protect them from cracking. Alternatively they are put into a mould and braced with strips of wood, cramps and packers. The two halves are overlapped and feathered. Ash and elm are especially used for this part of the wheel because of their natural springiness. The turnery of the spokes is kept very simple, in order not to interfere with the spinner's hand pushing the wheel. The number of spokes can vary from six to fourteen, and is sometimes an odd number, which gives added strength since no two spokes are then in direct alignment at the hub.

The Faeroe Islands have a hoop rim spindle wheel of an unusual type. There the men did the wheel spinning while the women continued to make warp yarn with the spindle and distaff. These spinning wheels are built into the wooden walls of the farm-houses, the driving wheel itself being attached to a wooden block nailed to the wall (18). The spindle is a separate unit, resting on a bench below; the upright is set into a small block of wood which, for tensioning, can be moved to any part of the bench, where it is kept firmly in place by two heavy stones. The two bearings are made of plaited wool wound tightly round to make oblong tabs (similar to the straw tabs of France). They are pegged into place at

18 Great wheels built into a farmhouse in the Faeroe Islands. From an old photograph c.1897. *Photograph: Pitt Rivers Museum, University of Oxford*

the back of the upright and the spindle is pushed through on the wheel side.*

However, a photograph shows that their neighbours in Iceland had the usual type of free-standing hoop-rim spindle wheel to which they sat.

SPINDLE WHEELS IN THE BRITISH ISLES

Spindle wheels existing in the British Isles today are chiefly of the large type which since the sixteenth century, perhaps earlier, has been referred to as the great wheel. This is also the name given to it in Wales (where the Welsh word, *troell*, 'wheel', appears in documents after 1400 – J. G. Jenkins, 1969). The great wheel was considered an essential piece of household equipment as late as the 1880s, for Wales was always mainly a woollen-manufacturing country, and undoubtedly quantities of great wheels were made there. When a well-to-do farmer could afford to pay for superior craftsmanship, it would have been on the great wheel that the wheel maker would exert his skill and ingenuity, and some fine specimens still remain (19).†

*According to Hoffmann, two Shetlanders, a man and a woman, were taken to the Faeroe Islands in 1671 to teach the people there how to spin on a spindle wheel.

† From Rogers' *History of Agriculture and Prices in England* (*1583–1702*) we learn that in 1587 a spinning wheel cost eight pence; in 1616 two shillings and four pence, while Geraint Jenkins states that in eighteenth-century Wales the cost was about five shillings.

19 Welsh great wheel: wheel diameter 36in.,
length of bench 36in. (91·5cm.).
*Photograph: Welsh Folk Museum, St Fagans,
Cardiff*

In Ireland, which was always a sheep-producing and wool-spinning country, and where there seems to have been an abundance of poor people, the wheels are simple in the extreme and mostly have straw-plaited bearings. Both Horner and Born refer to the Irish spindle wheel as the long wheel, but Lillias Mitchell (1972) calls it the big wheel, and this is the expression that Irish women use today. Most of these wheels are worked from a standing position, as in Connemara, Mayo and the Aran Islands (where the wool is used for knitting), but in Kerry the spinner sits to her work (**20**).

In Scotland the name for the spindle wheel is muckle wheel, from the Lowland Scots for 'great'. In England the spindle wheel had many names; in the sixteenth century, as well as the great wheel it was called the wool wheel. This is the name which it took with it to North America and which is still commonly used there.

20 Irish spinner using a big wheel. From an
old photograph c.1905

Another familiar English name is the walking wheel, on account of the need
to step forward to the tip of the spindle and back again for the draw-out. It is
said that spinners who worked for the textile industry in Yorkshire and Lancashire walked the equivalent of 30 miles a week spinning wool. This may well have
been possible when it is remembered that children learnt to spin at an early age,
and with the large wheel it must have been an energetic business. Some spindle
wheels can be found with pairs of holes at intervals on the tilting spindle upright
by which the height of the spindle can be adjusted. Although this could be a form
of tensioner the spindle might also have been adjusted according to the height of
the spinner.

The Jersey wheel was yet another name given to the large spindle wheel, but
it appears to be only indirectly connected with the Channel Island of that name.
In the sixteenth century 'Jersey stockings, which were still finer than ordinary

worsted were originally a speciality of the islands of Jersey and Guernsey' (Joan Thirsk, 1973). Yet for some unexplained reason the word came to be used in Lancashire. Lewis Paul in his patent of 1738 refers to combed wool as jersey, and as late as 1882 Beck's *Draper's Dictionary* states 'Jarsey is still the local name for worsted in Lancashire'. Although the great wheel is principally connected with the making of woollen yarn, when worsteds were made in the North it would seem that the great wheels were used for making the yarn and the wheel took the name from the yarn, since by that time jersey or jarsey had become a term loosely used, meaning worsted whether for the cloth or in the knitted stocking industry. According to J. James (*History of the Worsted Manufacture*, 1857) the Jarsey man took worsted yarn from Lancashire to Yorkshire.

Great wheels that have survived to the present day no longer have their benches resting on the floor at the spindle end (as shown in eighteenth-century engravings) but have three or four legs to keep them level. It may be that such wheels have lasted because they were made for individuals and were treasured possessions, used only domestically, since it is much less tiring to have the spindle raised to a convenient height enabling the spinner to work standing straight instead of with endless bending. Whether those who had owned two-legged wheels added a further leg under the spindle, or whether they were happy to break them up for firewood once they were no longer a means of livelihood, we may never know.

Though most great wheels are quite simple in their construction, from time to time a more elaborate arrangement can be found. At Snowshill Manor is such an example (21). The spindle and whorl are mounted on a bat head at the top of a short post, all of which is removable. The uprights supporting this are pivoted into their slots in the bench by a wooden pin so that they can tilt when the tensioner, which passes through a small turned post and into a cross bar between the uprights, is tightened. The two short pillars on the bench prevent a pile of rolags from slipping along and getting entangled with the driving band. The far edge of the bench has deep-cut scalloping, a decoration often found on all types of spinning wheel, and a single turned post supports the hoop rim wheel. This magnificent wheel might well date from the late eighteenth or early nineteenth century, and it appears from the records of the collector (the late Mr Wade) that it came from Aberdovey in Wales.

It is usually accepted that smaller wheels at which the spinner sat were used for cotton spinning (although large ones were also probably used). Pictures of English cotton spinners show them using a wheel which has a diameter of approximately 30in. (76cm.), and it is noticeable that the spindle is attached to the flat upright on the side away from the wheel, so a small hole is cut in the upright to allow the driving band to reach the whorl. Few of these spindle wheels remain today (even the well-known cotton manufacturers, Horrockes Ltd, who had kept one as a souvenir of the past as late as 1958, have it no longer) but one or two spindle wheels with spindles bolted on to the uprights with metal

21 Detail of a great wheel showing more elaborate turnery; wheel diameter 36in. (91·5cm.). The four grooves on the whorl are of equal diameter.
Photograph: The National Trust, Snowshill Manor, Gloucestershire

brackets, and with very small whorls, may be relics of cotton wheels.

It is perfectly possible to spin flax or hemp on a spindle wheel, and one hears of instances when this was done to perfection, but the long fibres are more difficult to manage when held in the hand, so they were tied round the spinner's waist. By and large, however, spinning in this fashion was for coarser types of yarn, and although rope making with its various methods of twisting is beyond the scope of this book it is worth mentioning the spinners who made twine for fishing lines and nets. It is known that from at least the beginning of the thirteenth century Bridport in Dorset was the centre of this industry in England. The raw material was grown locally and as late as 1890 twine was still spun by hand. The wheel was a large one (an example in the Bridport museum has a diameter of $33\frac{1}{2}$in., 85cm.), not unlike the Chinese multi-spindle wheel, with several rotating hooks (known as cogs) mounted in a double wooden arc above it. Between the two arcs was a whorl, attached to each cog, with four grooves for the driving bands. There was no treadle, but instead a pole 48in. (122cm.) long was attached to the wheel at the rim edge between two spokes, while the other end fitted loosely into a hole in an upright board or post a little distance away. A child was employed to turn the wheel by means of this pole from one side while the spinners walked backwards from the rotating hooks on the other side, paying out fibres with each hand as they went and so making two yarns at a time which, on the return journey, were twisted together. The twine was supported as it was spun by wooden skirders (resembling upturned wooden rakes) attached to trees or mounted on stands (22). These were indeed walking wheels, and spinning walks, ways, 'wyes' or more familiarly, rope walks were a feature of many fishing ports and could be anything from 50 to 1,200 yards in length. In Bridport the work was done by women but in Great Yarmouth, Norfolk, the twine spinners were men. A Yarmouth newspaper cutting of December 1928 describes the spinner 'deftly releasing hemp from around him in sufficient quantity through a piece of cloth or duffell', which would seem a most necessary precaution against cut hands. 'The spinner upon his return journey' continues the article 'to finally make the twine used a "top" (a grooved conical-shaped wooden top), this binds the threads into one twine.'

Horner described a similar type of spinning in Montrose, Scotland, which he

22 Women spinning hemp for fishing nets. One of four pictures depicting the netting industry painted and presented to the Bridport Town Hall by Fr. H. Newbury (1855–1946) in 1925. (He claimed to have taken great trouble in recording the scene accurately, having lived in Bridport as a boy when hand methods were still used.)
Photograph: Science Museum, London

said was a Dutch system, and Born states that hemp was spun on spindle wheels (probably of this type) until the early 1930s in Holland.

Accelerating the spindle

The principle of transferring the drive from one wheel to another before it reaches the spindle and thereby increasing its speed can be met with in India, where one is told it has ancient origins. This type of Charka has two disc-shaped driving wheels which rotate in a horizontal plane. The only example I have seen is contained in an oblong wooden box 23 × 8in. (58 × 20cm.). The larger of the two wheels, $5\frac{3}{4}$in. (14·6cm.) in diameter, is on the spinner's right and has a knob on it for turning. This wheel has a grooved circumference and is linked to the hub (or pulley, $1\frac{1}{2}$in., 4cm.) of the smaller wheel with a single driving band. A second driving band passes round the similar groove of the smaller wheel

(diameter 5in., 13cm.) and links it to the spindle whorl ($\frac{1}{4}$in. diameter, 6mm.). The delicate steel spindle, with a metal disc at its base for building the cone of yarn against, is supported between two uprights and the whorl is held centrally between them by two pieces of tubular coloured glass threaded on to the spindle, one either side. The spindle rotates very fast and is really only suitable for its original purpose of spinning the short fibres of cotton. Gandhi, when in the 1920s he advocated hand spinning of cotton as a means of relieving India's poverty and also because he believed that spinning gave peace of mind, had a Charka made to his specification on similar lines to that just described. Such wheels are made in India today (23). The spindle uprights are collapsible, and the box hinged in the middle with the wheels in the lid, so that it can be closed up and carried about and a person need never be without a spinning wheel.

In North America wool wheels in general appear to have larger wheel diameters than those in Britain, with an example from Canada (now in England) reaching 48in. (122cm.) diameter, and an overall height of 60in. (152cm.). A single turned post supports the wheel and, echoing its turnery, a single post holds the spindle, often a separate unit fitted into the top of the post. Therefore, when in 1810 Amos Minor, of Marcellus, New York, started to manufacture an accelerating head which he had patented in 1803, it could be fitted into the top of the spindle post. The smaller wheel, approximately 5in. (13cm.) in diameter, was made as a unit with the spindle and the driving band linking the two was quite short. The large wheel drives the hub of the smaller one. This unit is sometimes referred to as a Minor's head, but a number of firms manufactured it, mainly in New Hampshire, and it is doubtful whether Minor always received his patent royalty.

23 A modern Charka made in India on lines specified by Gandhi. The spindle folds away when the box is closed; larger wheel diameter 5in. (13cm.), smaller wheel diameter 3in. (7·6cm.). The batten attached to the box is for the spinner to sit on to keep it steady.

24 Hurdy wheel from Nova Scotia, Canada, 1860–1880, with a mass-produced accelerating spindle head. Wheel diameter 17¾in. (45cm.). *Photograph: Royal Ontario Museum, Toronto, Canada*

In Canada, the Minor's head was widely used on spindle wheels. It was also added to a simple construction consisting of a wheel mounted at the end of a plank of wood which was clamped to a table. This wheel, used particularly for plying, was called the hurdy wheel* **(24)** (H. D. & D. K. Burnham 1973).

It will be noted that this is one of the rare instances where the wheel, which has a handle on one of the spokes, is turned by the left hand and the yarns controlled by the right. Apparently the older people of the Maritime Provinces say that plying heavy yarns throws a spinning wheel out of alignment. In fact, the advantage of plying on the hurdy wheel, or for that matter on any spindle wheel, is that it puts no restriction on the size of the yarn, whereas on the later spinning wheel this is restricted by the size of an orifice (see the description of a flyer wheel in the next chapter).

Spindle wheels, when used for winding bobbins or spools, sometimes have an accelerating wheel arrangement for increasing the speed. They can have two wheels of equal size, one above the other, or a small wheel on the bench driven by a larger one.

WINDERS

Bobbin winding was a job done either by a small boy or a woman. A weaver can use up a bobbin in the same time as it takes to wind one.

*From the Scots hurdy-gurdy, a name given to various things turned continuously by hand, such as the musical instrument.

'The weaver next doth warp and weave the Chain,
Whilst Puss his cat stands mewing for a skeine
But he laborour with his hand and heels
Forgets his and cries "come boy with queles".'
(R. Watts, 1641)

There are many kinds of winders, but broadly speaking there are the open ended types and the closed end types, the former being the ones that can also be used for spinning, plying or silk throwing. The wheels were usually of medium size, about 18–20in. (45·5–51cm.) in diameter, with a handle on the axle, but large wheels were also used and some were probably dual purpose. The expression 'bobbing' or 'bobbin' was used for spindle wheels when used for winding. J. Lawson in his letters on the *Progress of Pudsey* in the nineteenth century wrote: the weft yarn 'is still wound on to bobbins on the one spindled wheel ready for the weavers shuttle . . . and we have seen the winders who are mostly women, girls and boys, with the blood dropping off their fingers caused by the friction of the yarn in winding these bobbins'. (They did not seem to have used a protective piece of duffle like the Yarmouth twine spinners.) Tim Feather, the last of the hand-loom weavers in Keighley, Yorkshire, who died in 1910, used a great wheel (with a diameter of 37½in., 95cm.) for winding his bobbins. (The wheel is now in the Cliff Castle Museum, Keighley.)

Winders often had a shallow boxed arrangement on the stock around the spindle support to hold the spare bobbins. When the spindle was enclosed at both ends it fitted into holes or slots across a deeper box. The enclosed type was also used for winding the warp yarn on to larger bobbins or spools with raised ends. From seventeenth-century Dutch paintings we see that this was a job sometimes given to old men. Once wound, the bobbins were transferred to creels (racks) for making the warp. (One imagines that the woman in picture vi of plate 7 is winding warp yarn rather than weft, since the wheel is beside the warping mill and not a loom, but of course one cannot be sure. From the position of her hands, always the most reliable way of distinguishing between spinning and winding, and with the swift beside her she is certainly doing winding of some sort.)

A simple type of winder consisting of a single rod with a small wheel half-way along and supported over a long narrow box, is depicted in the painting 'Return of Odysseus' by Pintorichio of Perugia (1434–1513), now in the National Gallery, London. The same type, in no way changed, was used in Italy until quite recently, but Horner collected from Tenerife in the Canary Islands a similarly designed all-metal implement, mounted on a log of wood but with the uprights placed so that the rod protruded at one end, and it was apparently used for spinning. Italy, too, had the spindle mounted in this way on a flat board but it was only used for winding. In both Sicily and Greece the same tool was placed vertically in a recess at one end of a board on which the person winding sat. The bobbin was tied to the small wheel at the top and, to wind, the post was vigorously twirled.

Chapter Three

The flyer spinning wheel

'Le roüet est une machine qui nous paroît simple & qui, exposée par-tout à nos yeux, n'arrête pas un instant notre attention, mais qui n'en est pas moins ingénieuse.'*

Diderot, **Encyclopédie**. *'Fil'. 1756*

A great step forward in the evolution of the spinning wheel was the addition of the flyer, a U-shaped piece of wood or metal attached to the spindle, and two whorls, one also attached to the spindle, and the other forming part of the bobbin and having a smaller circumference. This made it possible for the spun yarn to be wound on to the bobbin simultaneously with the actual spinning, and thus over-came the intermittent process of the spindle wheel and made spinning a con-tinuous operation.

The spinning wheel with flyer is sometimes referred to as the long fibre wheel, but more frequently as the flax wheel, since making a linen yarn was one of its main functions for around 400 years. The layout of the early flyer spinning wheels was similar to that of the spindle wheel in that the spindle with its flyer was mounted horizontally beside the wheel. In the strict sense of the word these spinning wheels are not always horizontal since the benches often slope, but the term is used to distinguish them from the vertical spinning wheels, where the spindle and flyer are mounted above the wheel. On both types the wheel rotates in a vertical plane.

Spinning wheels are mostly constructed of wood, with metal for the spindle and wheel axle, and leather for the bearings. Many of the hardwoods used for turnery and furniture making are found in spinning wheels, including beech, oak and box, yew, sycamore, elm, ash, chestnut and some fruit tree woods such as cherry, apple and pear. Pine and birch were also used. In the eighteenth century when it became fashionable to own a beautiful spinning wheel, the more exotic woods such as mahogany and satinwood were used. Thick layers of stain or varnish can often make it difficult to identify the wood but one commonly finds more than one type used in the making of a spinning wheel.

*'The spinning wheel is a machine which appears to us simple and which, seen everywhere, does not for an instant hold our attention, but which is none the less ingenious for that.'

Before attempting to trace the evolution of the horizontal spinning wheel, it will be best to look at the individual parts as we know them today (25).

THE STOCK

The stock (1), table or bed, is the bench of the spinning wheel. It is a sturdy block of wood varying from 2–5in. (5–13cm.) thick and can be quite plain or decoratively carved. It is usually oblong but it sometimes tapers, getting larger towards the wheel end. The stock can be horizontal or slope downward towards the wheel at an angle of as much as 60° to the ground. Sometimes the stock has in it a small covered recess to hold a spare bobbin or just a round hole in which a bobbin can be placed. Recesses smaller than a bobbin's length were presumably used for odds and ends and the threading hook (the use of which is explained later). The stock is supported by three legs, two underneath the wheel and one under the spindle and flyer (occasionally there are four legs). Many spinning wheels have a hole in the stock for holding the distaff (2). The hole is usually on the left, sometimes in the back corner, sometimes in front, while some spinning wheels have both, for the spinner to choose the most convenient according to the length of her fibre and the way she spins it. A distaff may include an upright and an arm by which the distaff itself, with the fibres attached, can be moved into the desired position. The actual distaff must be easily removable for attaching the fibres and for this reason distaffs are seldom screwed into place but have tapered ends which slip into holes.

THE DRIVING WHEEL

The driving wheel (3), the most prominent feature, is mounted on the right-hand side of the stock. While its function is to drive the spindle and bobbin, it also plays a major part in the aesthetic look and character of each individual spinning wheel. It is made in a variety of sizes ranging from about 7in. to 26in. (18–66cm.). The number of spokes can be from six upwards (only occasionally four) and seldom more than 20. The hub is a round block of wood, 2–4in. (5–10cm.) thick and about the same in diameter. The axle is composed of either a metal shaft through the hub, or two short metal rods, one inserted on either side of the hub. It is important that these be in dead alignment – one of the factors upon which the true running of a wheel depends. The rim consists of four, six or eight sections or felloes, cut with the grain running tangentially to the curve. The depth of the rim varies but is seldom less than $1\frac{1}{4}$in. (3·2cm.); the outer edges are raised sufficiently to prevent the driving band from slipping off and the circumference is sometimes grooved for the band. The spokes are tenoned into the hub and the felloes are drilled on the inside for fastening over the tapered ends of the spokes while they are joined together with dowells or tongues. This resembles the wheelwright's method, but in an alternative way frequently used on spinning wheels the spokes are cut to the exact inner circumference of the joined felloes, and are pinned in from the outer rim with headless nails; or else there is

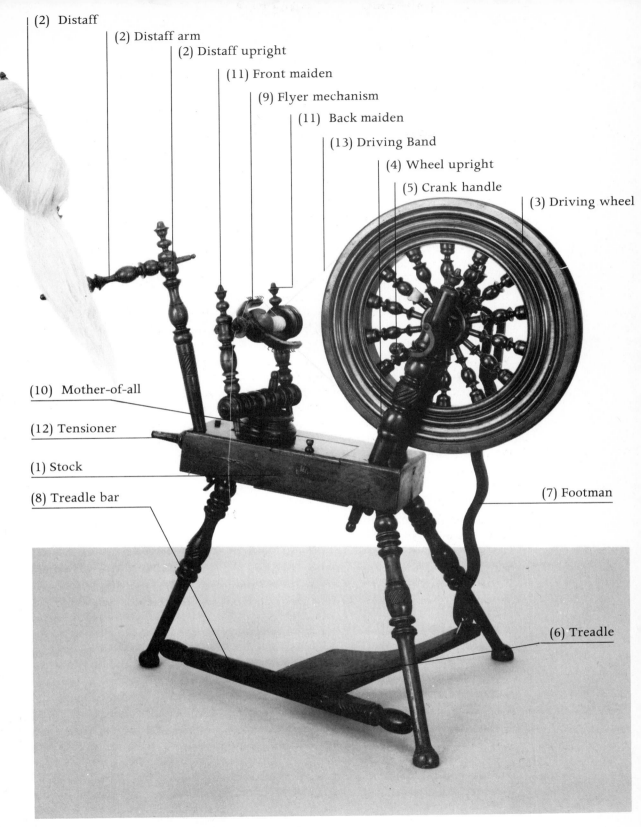

(2) Distaff

(2) Distaff arm

(2) Distaff upright

(11) Front maiden

(9) Flyer mechanism

(11) Back maiden

(13) Driving Band

(4) Wheel upright

(5) Crank handle

(3) Driving wheel

(10) Mother-of-all

(12) Tensioner

(1) Stock

(8) Treadle bar

(7) Footman

(6) Treadle

25 Spinning wheel with flyer and treadle (horizontal shape), probably late eighteenth century, wheel diameter 15½in. (39·5cm.).

Photograph: Crown Copyright, Science Museum, London

a rebate, a shallow lip, on one side of the inner rim so that the headless nails can be hammered into it at an angle through the spokes. Sometimes the ends of the spokes are cut to fit over the rebate and then nailed. Occasionally the wheel is dished, that is, the spokes slope at a slight angle away from the hub. This was a method used for cartwheels bearing heavy weights on uneven surfaces and is not necessary on a spinning wheel. It does make it easier to slip the circle of felloes over the spokes, but equally likely it occurs simply through wheelwrights' tradition. If the wheel is made of four felloes, the two opposites are larger than the others. Some wheels are made without felloes, but from three pieces of wood joined edge to edge with the grain running in the same direction. The middle piece is the largest and has a hole in the centre for the hub and spokes, and two smaller pieces are added either side to complete the rim.

The wheel uprights (4) are two sturdy turned posts fitted into the stock and glued or pegged, sometimes both. The top of the uprights are slotted to take the axle, the slots sometimes being reinforced with brass or other metal to minimise wear which would upset the smooth running of the wheel. Again important, the two slots must be correctly aligned. The posts can be perpendicular to the stock, or sloped outwards to the right. Either the slope of the stock or the angle of the posts may bring the wheel beyond the end of the stock. In view of this the uprights are sometimes supported by stays.

To prevent the wheel from jumping out as it rotates, wooden pegs are fitted across the axle. Alternatively the pegs are fitted through wooden wedges, often with decoratively turned knobs, or a wooden cap is fitted to the top of each upright with its edge overlapping the axle. Each of these arrangements allows for easy access to the axle to lubricate it or for removal of the wheel should this ever be necessary. The back of the axle extends to bear a crank; in the spinning wheel illustrated the axle also extends forward to bear a curved crank (5) with a knob handle. Cranks can be either straight or curved, the latter being the more usual on older spinning wheels. The throw of the crank rarely exceeds $2\frac{1}{4}$in. (5.7cm.), giving a treadle rise of twice this, more than which could tire the spinner.

THE TREADLE

The treadle (6) is linked to the crank by the footman (7), also called link-rod or pitman. This has an oblong hole at the top to fit over the knob at the end of the crank, though on some modern spinning wheels the footman is attached to the crank by screw and lock nut. The footman is usually a wooden rod, sometimes made in a wavy shape as illustrated, or, instead, a piece of cord is used, often with a tab of leather at one end with a slot cut to fit over the end of the crank. Sometimes the footman passes through an opening cut in the stock, in which case the crank comes inside the back upright and is thus well protected.

A treadle bar (8), most usually with a metal pin either end, fits between the front two legs, from which position it pivots with the movement of the treadle. This is made of one or more pieces of wood, at least the width of the human foot, and reaches to meet the footman at the back. It can be nailed on to the treadle bar but it is equally usual to have leather hinges, in which case the treadle bar is a fixed one. The footman and the end of the treadle both have a hole through them and are connected by a leather thong, or often by a piece of string, in such a way that the two do not quite touch, thus forming a loose universal joint. Triangular treadles are often found: a bar from the treadle bar extends to the footman and a piece of wood is fastened across the two. Treadles can be carved in all sorts of shapes; such decorative work is usually on the left-hand side, since the spinner sits in front of the treadle and generally uses her right foot, which tends to wear down the wood on that side.

THE MOTHER-OF-ALL AND MAIDENS

The mechanism for twisting is mounted on the left-hand side of the stock. This is sometimes referred to as the spinning head but will here be called the flyer mechanism (9) to avoid possibility of confusion with the spindle of the spindle wheel. The flyer mechanism is supported upon the mother-of-all (10) – a homely name whose origins are lost in the past, though a suggestion is that it may be a corruption of the German *Mutter* in the sense of a nut. It normally consists of a round disc resting on the surface of the stock and of a turned cross bar attached above. Beneath the disc it extends to rest in a slot in the stock, sometimes held in place by a nut below. It is positioned so that the whorls of the bobbin and spindle are in strict alignment with the driving wheel.

The mother-of-all supports two turned uprights known as maidens (11) or sisters. The tops of the maidens provide an opportunity for elegant turnery and while matching each other, it is seldom that any two pairs are identical even when by the same maker. The purpose of the maidens is to hold the bearings which support the spindle. These are made of leather, the front bearing being a wedge with a hole in the middle while the back one may be the same or simply a loop of leather. Their distance apart depends on the length of the particular spindle. The back maiden is glued into position with the bearing on the right, which fits through the maiden and is pegged or glued into position. The front maiden has to be in some way movable to make it possible to release the spindle in order to remove the bobbin when necessary. By one method the maiden fits firmly, but not too tightly, into the mother-of-all, and is simply turned sufficiently for the end of the spindle to drop free. In other methods the maiden is made to slide forwards along the mother-of-all, a wooden nut having first been loosened. In this case either the mother-of-all or the maiden is slotted.

The maidens are always parallel with each other, but their angle to the stock varies. They can be absolutely upright or tilt away from the wheel at an angle of up to 45°.

TENSIONER

The tensioner (12) is a wooden screw with a handle at one end. The screw enters the left end of the stock and engages in the extension of the mother-of-all. When the handle is turned clockwise it draws the mother-of-all and flyer mechanism

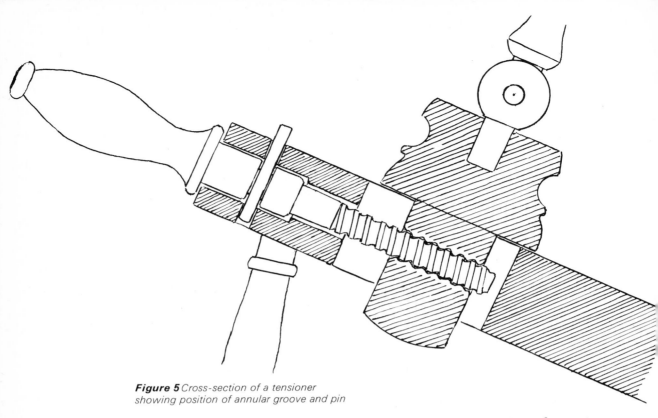

Figure 5 *Cross-section of a tensioner*
showing position of annular groove and pin

away from the driving wheel. The amount of movement is about $\frac{5}{8}$–2in. (1·6–5cm.). In the tensioner between the screw-thread and the handle is an annular groove. A pin (usually of wood but sometimes made decoratively in bone or ivory) enters from the top of the stock to engage in this groove, and this prevents the screw from coming out of its position (*figure* 5). Without the pin through the stock (and many spinning wheels are constructed without one) it can be difficult to move the mother-of-all towards the wheel without putting on pressure. Sometimes the mother-of-all runs on two wire guides to prevent it from swinging out of alignment.

FLYER MECHANISM

(*Figure* 6) The spindle itself (1) consists of a solid metal shaft which fits firmly into the centre of the U-shaped flyer (2), and is extended towards the spinner by a hollow shaft, approximately 1–2in. (2·5–5cm.) long. The shaft's open end is called the orifice (3) and there are two side openings, the eyes (4), one on either side of the shaft near the base of the flyer. On old wheels the diameter of the orifice averages about $\frac{3}{16}$in. (5mm.), but some modern ones have them larger for spinning a thick yarn. The end of the solid part of the shaft rests in the bearing of the back maiden, while the hollow shaft rests in the bearing of the front maiden with the eyes inside it and the orifice outside; some sort of ridge or collar (5) ensures that it does not slip out of the bearing. An inch or two from the end of the solid shaft is a screw-thread (6) (usually a left-handed one because the spindle is more often rotated clockwise). Screwed on to this is a grooved disc of wood, the spindle whorl (7). In this there are often two grooves, sometimes of the same circumference, but usually one is very slightly deeper than the other, for getting a tighter or looser twist, although it is the one that is less deep that is normally most used. The spindle shaft, whorl and flyer all rotate as one, and are generally, as here, referred to simply as the flyer.

The flyers themselves are generally made of wood and are of no uniform shape or size. Some have a deep curve fitting closely round the bobbin while others have a shallow curve making a wide span between the ends of the two arms. The shape of the U is largely dependent on the shape of the bobbin (or *vice versa*). There must be sufficient clearance between flyer arms and the bobbin a) to avoid friction b) to allow the bobbin to be filled to capacity c) so that the yarn cannot catch on the nearest raised edge of the bobbin. The flyer arms must end clear of the bobbin whorl to avoid entanglement with the driving band.

In Britain and more northerly parts of Europe the flyer is usually made from a single piece of wood, but further south, in Austria, northern Italy and Bavaria, the construction of the flyer is rather different. The hollow shaft is made of wood forming part of a bulbous centre, while the arms are separate shaped pieces of wood, often of flat section, attached to this centre. Sometimes the arms, though curved on the inner side, are angled on the outer, while others are quite straight, pointing outwards from the centre to make almost a V-shape.

On the arms of the flyer are a series of small hooks (8) for guiding the yarn on to the bobbin; to keep the yarn evenly distributed on it, the spinner is obliged to move the spun yarn from hook to hook. It will be noticed that each eye of the spindle shaft is in alignment with one of the flyer arms. The spun yarn passes first through the orifice and through one of the eyes on to the nearest flyer arm, and then along the hooks and on to the bobbin. It should not go from the eye to, say, the third hook, but must always lie along all the hooks, else it will catch on the bobbin. It is important that the first and last hooks are so placed that the bobbin is filled right to both ends.

The hooks are normally placed on the trailing side of each flyer arm, which is the upper left-hand side when Z twisting (with the flyer rotating clockwise). This prevents the thread from bearing unnecessarily hard on to the curve in the flyer arm and cutting into the wood. Even so flyers are found with cuts not only at this point but also where the thread passes round the hooks, when formerly these were merely pieces of bent wire; the yarn then dragged over the wood. These cuts can be damaging to yarn subsequently spun, particularly if the cuts are deep from spinning flax and a lightly twisted wool yarn is then made, since this may catch in the grooves. Nowadays the hooks are usually in a small cup-hook shape and therefore the yarn is lifted by the curve of the hook slightly above the flyer arm. It is important that the metal is not too soft since it is even possible for the hooks to become grooved. Some flyers have a short metal strip on the curve of the arm to protect the wood at this point from the rubbing of the thread as it passes on its way to the first hook.

When the hooks are placed on the upper right-hand side it is for yarn that is to be spun in an S direction, with the flyer rotating anticlockwise. For reasons stated on page 18 (chapter I) flax is often spun with an S twist and it can be an indication that a wheel was used for flax if the hooks are found in this arrangement. It is convenient if both sets of hooks are on the same side, in which case only one eye is needed, facing centrally, and the flyer can then trail the thread in either direction. This is also useful for plying, since normally yarns are twisted together in the opposite direction from which they were spun.

Instead of hooks, there are sometimes holes in the flyer arms, and a single peg, thorn or metal eyelet is moved from hole to hole to guide the yarn on to the bobbin. This is found more in Germany, Austria and adjacent countries, but occasionally British flyers can be found where such holes have been filled in and hooks subsequently inserted. (In the eighteenth century the mention of quills and feathers in connection with flyers may indicate that at one time these were used instead of hooks.) When the flyer is made of brass or other metal, slits and round holes are cut out along one side of one or both of the flyer arms instead of hooks. It is interesting to note how narrow the slits usually are; they give an indication of the fineness of the yarn that was spun.

THE BOBBIN

(*figure* 6) The bobbin (9) is a totally separate unit from the spindle and flyer. It is normally of wood, with a grooved whorl (10) at one end and at the other end a disc which is flat on the inside and is slightly domed on the outside to fit into the U of the flyer. (Some bobbins made for use with the more elegant spinning wheels have ends made of ivory.) The stem is hollow and should fit loosely on to the

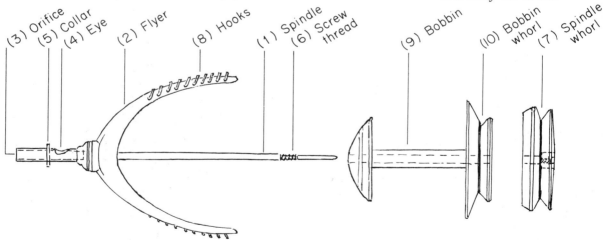

Figure 6 *Flyer mechanism: flyer and spindle;*
bobbin; spindle whorl. (1) spindle (2) flyer
(3) orifice (4) eye (5) collar (6) screw-thread
(7) spindle whorl (8) hooks (9) bobbin
(10) bobbin whorl

spindle shaft since it must move independently. However, it should not be possible for the bobbin to shift along the spindle shaft since this could cause the bobbin whorl to guide the driving band too close to the tips of the flyer arms and catch in the last hook; if this occurs it can be easily rectified with a washer placed on the spindle before the bobbin.

To take the bobbin off the spindle or put one on to it, the spindle whorl is first screwed off. When the two whorls are in position they lie close together. The measurements of their circumferences differ from one spinning wheel to the next, but the relative difference can vary from ½in. to about 4in. (1·3–10cm.). Bobbins are made to go with particular spinning wheels and are seldom interchangeable unless by luck. Most wheels are supplied with three or four bobbins.

DIFFERENT TYPES OF DRIVE

A driving band (13 in plate 25) links the driving wheel to the flyer mechanism and this can be arranged in three different ways.

1   Bobbin lead (26)
(or doubled band drive which will be used in this book to differentiate from *No.* 2).

A single continuous band goes twice round the driving wheel, once round the spindle whorl and once round the bobbin whorl. To the uninitiated this looks like two bands but on looking carefully it will be seen that the band crosses itself, so it is a doubled band. (This is not visible in the illustration since the

26 Detail of the flyer mechanism showing a bobbin lead, doubled band drive.

Photograph: Crown Copyright, Science Museum, London

cross-over occurs only a few inches from the wheel.) If the driving wheel is being turned clockwise the cross will appear between the bottom of the wheel and the two whorls; to avoid unnecessary friction, the band coming from the spindle whorl should be underneath. If the driving wheel is turned anti-clockwise the cross will appear between the top of the wheel and the two whorls; the band coming from the bobbin whorl should be underneath. The cross will find its place automatically according to the direction of the rotation of the driving wheel when it is turned and the band is sufficiently tightened to rotate both the spindle and the bobbin. Each rotation of the spindle puts one twist into the forming yarn, but at the same time, since the bobbin is rotating faster owing to its smaller circumference it is drawing-in the yarn on to itself, the flyer acting as a guide. The difference between the circumferences of the two whorls affects the ratio of twist to draw-in. If this difference is very small, say $\frac{1}{2}$in. (1.3cm), the draw-in will be very slow and therefore the yarn will be given a lot of twist, often too much. It is better to have a greater difference between the whorl circumferences, averaging 3in. (7.5cm.), since the spinner can always resist the draw-in to give the yarn more twist before letting it wind on to the bobbin.*

It will be realised that for the spinner to control the draw-in and therefore the amount of twist, the driving band slips on the bobbin whorls. If the band is so tight that it cannot slip the yarn will be drawn in too fast, before it has sufficient twist to form a coherent thread, and it will most likely disintegrate. But if the band is too loose, the draw-in will be too slow, resulting in an over-twisted yarn.

As the bobbin fills, its circumference becomes greater, so that more spun yarn is needed to complete one turn of the bobbin. There would then be fewer twists per inch on the thread and it would therefore become increasingly soft. The bobbin is also getting heavier, which slows it down, reducing the difference in the movement between the two whorls. Therefore, to keep the ratio of twist to draw-in constant, the tensioner is given a quarter-turn at regular intervals during the filling of the bobbin to tighten the band by moving the whorls slightly further away from the driving wheel.

2 Flyer drag (27)
(or bobbin lead, but not to be confused with *No.* 1; it is also sometimes called 'spool driven').

The driving wheel is linked to the groove of the bobbin by a single driving band. There is no spindle whorl, since in this instance the flyer is dragged round by the yarn. As spinning commences and there is some tension on the thread, the differential speeds will come about without any further assistance. However, to have the necessary control on the flyer as the bobbin fills with yarn and its diameter increases, a short friction band (drag) is required. It is placed across the

---

*The difference in the circumferences is an important point to look for when buying a spinning wheel. The whorls can be measured quickly by placing a piece of fine cotton round each groove and so comparing them.

27 Detail of a German spinning wheel showing a flyer drag with friction band across the hollow shaft and the flyer with holes instead of hooks.
Photograph: Ulster Museum, Belfast

hollow shaft of the flyer and can be tightened by a peg. In effect, this changes the shape of the front maiden, the top of which is in this case shaped to hold the hollow shaft. Often a piece of leather or cloth, such as felt, rests over the top of the shaft and the band is tightened across this. Sometimes the back maiden is similarly shaped and there is a second friction band across that end of the spindle. In plate 27 the maidens are placed horizontally not vertically. This type of drag is found more frequently in Germany and Central Europe with the kinds of flyers described above as coming from these countries.*

*A drag on the flyer can now and then be found in conjunction with the doubled band, for they both give a bobbin lead. In this case there is a whorl on the spindle as well as on the bobbin. Examples where this occurs, and yet there is but one groove in the driving wheel, may represent adaptation from a flyer drag. The conjunction of flyer drag and doubled band makes possible even finer control of the flyer, and may have been particularly necessary when there was little difference in the circumferences of the whorls. It has been suggested that extra drag on the flyer can be given (when it is a doubled band drive) not by a friction band, but by turning the front maiden so that its bearing presses against the collar of the spindle; this was no doubt in fact done, but it does not give as satisfactory a control as a friction band.

28 Detail of a spinning wheel from the
Shetland Isles showing a bobbin drag with
friction band

3 Bobbin drag (28) (or flyer lead)

Another method of achieving the required difference in speed between spindle
and bobbin is to reverse the process, driving only the flyer. This uses a single
driving band which links the driving wheel and spindle whorl and must be
tight enough to turn the spindle and flyer without making it difficult to treadle.
As there is no drive to the bobbin, this rotates without control until spinning
commences, being then dragged round by the yarn, though only at the speed of
the flyer, and there would be no draw-in or wind-on of the yarn. Therefore a
brake is required on the bobbin. A friction band is attached to some part of an
upright or the mother-of-all, and is then taken over the bobbin whorl and attached
to a turnable peg on the other side. There is no need for the drag to touch more
than about a quarter of the circumference of the whorl. The friction band is
tightened just enough to make the bobbin rotate slower than the flyer and wind

on the spun yarn. In this manner the two whorls are tensioned independently. When it becomes necessary to counteract the growing diameter of the bobbin as the spun yarn is wound on, a slight turn of the peg is made. This is a very sensitive adjustment, which some spinners prefer since they feel they have a finer control. Whichever method of breaking is used the ratio between the twist given to the yarn and the speed of wind-on must at all times be constant.

Since in this case the flyer, rotating the faster, is winding the yarn on to the bobbin, when the yarn is wound off, the bobbin will turn in the same direction as when spinning. On the other hand, with the two previous drives, where the bobbin winds the yarn on to itself (the bobbin moving faster), when the yarn is wound off, the bobbin will rotate in the opposite direction to the wind-on. With all these different drives, the flyer is rotating in the same direction as the wheel.

An arrangement for the bobbin drag often found on Scottish spinning wheels is a small stick or piece of string joining the tops of the two maidens. One end of the friction band is attached to this string or stick (as shown in **28**). With bobbin drag wheels it is not essential to have a tensioner (though it is usually included) since there is no need to alter the driving band (unless, of course, it stretches) once it is tight enough to turn the spindle whorl and the differential speeds can be adjusted by the friction band. Quite often the bobbin drag is found as an alternative method of braking to the doubled band, on one and the same wheel, and then there has to be a tensioner.

The slower movement of the bobbin gives fractionally more time for the hands of the spinner to do their work, and therefore very slightly alters the rhythm of the spinning. This can make it a little easier for the beginner, and can also produce softer yarns.[*]

THE DRIVING BAND AND THREADING

The driving band has been made of different materials, gut or leather for instance, but if made of cotton, wool or linen it needed to be plyed for strength. Nowadays bands of man-made fibres are sometimes used and are satisfactory if they do not stretch, but many spinners still prefer to use a strong but not too thick plyed cotton or linen.

To measure the length of a doubled band, if there is no old one as a guide, the tensioner is turned to position the mother-of-all nearly as close to the wheel as it will go, and the band is wound round whorl and driving wheel once or twice according to the method of drive. The two ends are joined by overlapping or

[*]The popular modern New Zealand spinning wheel, the Ashford, works on the bobbin drag principle, but the arrangement is different from older wheels. The spindle whorl is placed at the base of the flyer between the orifice and the eye while the bobbin whorl is at the opposite end of the spindle, so that the whorls do not lie close together. There are two separate tensioners, and it is the back maiden which is turned to release the bobbin, making it unnecessary to alter the tension of the driving band.

splicing and then sewn, although many spinners tie a reef-knot and trim the ends. It is sometimes a good idea to use a knot in the first instance in case the driving band stretches when it is first used, but the material of the band should be sufficiently non-elastic not to continue to do so. Once the band is sewn it is not possible to remove it from the driving wheel without removing the wheel from its sockets, an ill-advised practice.

Friction bands need to be the length from the fixed point to the peg plus sufficient extra length for turning at least two or three times round the peg.

When starting from an empty bobbin one ties a length of spun yarn twisted in the same direction as required in the spinning (either S or Z) on to the stem of the bobbin and winds it round a few times; this could be termed a starter thread. The spare end is guided from the bobbin along the hooks of the flyer and threaded through the eye and the orifice. The need to thread it may be the reason for this end of the spindle sometimes being called the needle end, even though it is the other end that is the more pointed. A threading hook or bent piece of wire can be pushed into the orifice and out of the eye to catch the end of the yarn and pull it through, but some spinners prefer to push it through with their fingers. To place a bobbin on the spindle or remove it when full, the mother-of-all should be moved by the tensioner as near as possible to the wheel so that the band is slack and can be slipped off the whorls before moving the front maiden to release the spindle.

EARLY SPINNING WHEELS WITH FLYER IN EUROPE

Just as there is no positive proof as to where the spindle wheel first made its appearance, so it is unclear where the first spinning wheel utilising a flyer was made and who made it. It is evident that it is something that was evolved through experiment, and this could have occurred in more than one place. The need for an improvement that could spin long fibres of flax and long combed wool for warp, to replace the slow but reliable spindle and distaff used for this purpose, must have been a very conscious one. The most obvious places for such a development would be the textile centres themselves, although the spinning was done in wide areas all around, and it is likely that the silk throwing machines lay behind it.

There seems little doubt that the mechanisation of silk reeling and throwing was established in Italy at an early date, most probably first at Lucca (*Ciba Review* no. 80). There they tried to conceal the secrets of the machine's construction, but an exile from Lucca named Borghesano set up a water-driven throwing mill in Bologna in about 1272. Other exiles introduced the machines into Florence and Venice in the mid-fourteenth century. It was not until 1607 that information about throwing mills became available. This was in a book, *Nuovo Teatro di Macchine et Edificii* by Vittori Zonca. It is assumed that the water-driven 'Piemontese' twisting mill (**29**) discussed in this work was similar to the one introduced in Bologna and in use in Lucca.

29 Detail of the silk throwing machine
illustrated by Zonca in 1607 showing the
S-shaped flyers placed on top of the bobbins
and the reels above

It was first necessary to wind (without twist) on to a single bobbin the number of reeled silk filaments to be twisted. This could be done by using a spindle wheel. The bobbin was then placed on a spindle in the machine on top of which fitted an S-shaped wire flyer. As the bobbins rotated the threads unwound and received twist by passing through the two eyes, one at each end of the flyer, as they were drawn off the bobbin and wound on to a rotating reel above, the flyer and the reel moving slower than the spindle and the bobbin holding the filaments. There were thus two different speeds, and from this may have come the idea for the flyer spinning wheel.

The first silk twisting mill to be introduced into England was established in Derby about 1717–1720 by John Lombe, who is reputed to have disguised himself as a workman in an Italian mill in order to understand the workings of the machine. When Daniel Defoe visited Derby in 1724 he described the mill as follows: 'A curiosity trade worth observing as being the only one of its kind in England, namely, a throwing or throwster's mill which performs by a wheel turned by water [of the river Derwent]; and though it cannot perform the doubling part of a throwster's work, which can only be done by a handwheel, yet it turns the other work, and performs the labour of many hands.' It would therefore appear that it was still necessary to prepare the bobbins (what Defoe calls the doubling part, although the doubling is usually the actual twisting) on the spindle wheel.

It would seem that of the different spinning wheel drives described above the one which most resembles that of the silk throwing mill is the flyer drag, because the yarn is partially acting as a drag. If this were the earliest form of spinning wheel, there seems to be no evidence to prove it. The only hint could be that it is this drive which is found predominantly in the southern section of the total area associated with the flyer spinning wheel. No fifteenth- and sixteenth-century spinning wheels have apparently survived and therefore for that period we have to depend for our knowledge largely on pictures.

Italy was certainly one of the countries in which the flyer spinning wheel was developed since it was one of the many mechanical problems embraced by Leonardo da Vinci. Brought up in Florence (at a time when there was still prejudice against the spindle wheel for making warp thread) he must surely have been acquainted with the methods of the textile industry. His notes and drawings devoted to mechanisms for spinning, twisting and winding yarn are amongst the collection of papers known as 'Codex Atlanticus' and are dated to approximately 1490, by which time he was residing in Milan. As it is now known that flyer spinning wheels were in use before that date there can be little doubt that by experimenting with an existing mechanism he was not creating a new invention, particularly so as his interest in clock mechanism and cog wheels led him to something much more complicated and sophisticated than the simple arrangement of the hand-turned flyer spinning wheel. He used the same type of spindle and flyer for twisting, but he reversed the differential pulley action by giving the larger circumference to the bobbin whorl and the smaller to the spindle, a method that was not adopted. On the rare occasions when this method is found on a spinning wheel it will be noticed that the bobbin was probably not originally made for that particular wheel. As we have seen, the action of a bobbin moving slower than the spindle for winding on the yarn was in fact used on spinning wheels, but only with the aid of a friction band, which is not what Leonardo's sketches suggest. He also included a mechanism for winding the yarn evenly on to the bobbin and another for using more than one spindle at a time. In the former it was the spindle that oscillated while the bobbin remained stationary, but when the mechanism became used in a limited way on spinning wheels in the late eighteenth century the movement was reversed, and when introduced on to spinning machines it was also reversed and the spindles were placed vertically — not horizontally like Leonardo's.

The earliest record of a flyer wheel is a picture which appears in the Waldburg family's *Mittelalterliches Hausbuch* c. 1475–80, from Schloss Wolgegg near Lake Constance in the southern part of Germany (30). However, this is no evidence that it was invented there, nor is there any indication whether it had arrived there from the north or the south, but it is interesting to note that up to about 1450 Constance had been famous for its linens, 'Tela di Constanza' (*Ciba Review* no. 64). This fifteenth-century spinning wheel establishes the existence at that time of a small wheel (approximately 12–14in. diameter, 30·5–36cm.) — small compared with the Luttrell Psalter one for instance – driving, by means of a double band, a bobbin, spindle and flyer (just discernible along the flyer arms are several small pins or pegs). The flyer mechanism is supported between two uprights (maidens) with a cross bar underneath the spindle connecting them with a screw knob arrangement to release the spindle for replacing bobbins. Spare bobbins are visible on the lower shelf of the bench. The whole flyer mechanism fits into a slot cut in the extension of the top of the bench along which it can be moved, and it must be presumed that there is a wooden nut underneath to tighten it in position. Incorporated into the ornate bench is a cavity with a hinged flap and a ring handle, perhaps used for storing raw material. Decorative clusters of some sort are on the outer rim of the wheel, some to hide the joins with the spokes. The wheel is turned by a handle fixed to one of the spokes.

It is difficult to see from this drawing whether the drive is with a continuous band or two separate ones. If it is a continuous band, it certainly does not cross over on the lower level; nevertheless the two bands come very close together on the upper level and this could indicate an overlap; also, only one knot is visible, on top of the spoke with the handle. The wheel would be turning anticlockwise, giving an S twist. The distaff has an articulated arm and the vertical part would appear to come apart half-way up; the fibre looks like flax.

Nearly 50 years later, a similar spinning wheel appears in the Glockendon Bible, dated 1524. This bible, illustrated in Nuremberg for Ferdinand Albrecht the elder, is preserved in the Herzog August Bibliothek, Wolfenbüttel, Lower

30 Drawing of the spinning wheel in the
Mittelalterliches Hausbuch c.1480.
Photograph: Science Museum, London

Saxony. The miniature (in volume II, p. 1963) might well record a family scene
or event, since the lady, sitting at the spinning wheel in a fashionable pink hat,
her blue skirt covered by a large white apron, receives a letter from a man-
servant who bows and touches his hat as he hands it to her. Her two companions,
either daughters or maidservants, with bands round their hair, and simply
dressed, use a spindle and a free-standing distaff on a small round base. A large
red cat completes the domestic scene. The design of the spinning wheel differs
little from that in the *Hausbuch*, but owing to the small size of the picture the
flyer is not discernible, yet the bobbin is clearly between two uprights or
maidens, so the flyer can be assumed and also a tensioner, since the flyer mechanism is positioned on an extension of the bench top. This bench has the same
decoratively carved arches as the one illustrated in the *Hausbuch* and an articulated arm to hold the distaff. The fibre on the distaff again resembles flax, which
is bound round with a green band and gold cord to hold it in place. A similar
spinning wheel is illustrated in a Strasbourg calendar early in the sixteenth
century, the extended bench supporting the wheel and flyer mechanism placed
on a decorative chest. The woman is spinning from a distaff with her left hand
(illustrated in *Ciba Review* no. 22).

Meanwhile Flemish and Dutch artists of the first half of the sixteenth century,
with their precise techniques and eye for detail, have contributed valuable facts
to our knowledge of the early spinning wheels. There are noticeable differences
between the spinning wheels just described and those from the Low Countries.

31 A young woman spinning, 1513. Engraving
by Lucas van Leyden (1494?–1533).
Photograph: Cliché des Musées Nationaux

Gone are the solid gothic-looking benches; instead a small and more lightly
constructed table or stock is supported by legs. This is shown in an engraving by
the Dutch artist Lucas van Leyden dated 1513 (**31**). The most natural and simple

progression from the spindle wheel to a spinning wheel incorporating a flyer would be to slip the flyer on to the end of the protruding spindle. This is exactly what has been done in this instance. There is a single driving band only and no sign of a tensioner, but it is likely that a friction band would be across the bobbin whorl, for without this one cannot understand how the differential speeds could have been achieved. (Careful scrutiny shows a hint of a friction band on the front maiden, to which it would be attached.) The angle at which the woman sits to spin in relation to the flyer is rather curious, but it does give a very clear picture of the spinning wheel and this could have been the artist's intention. It is also possible that the spindle was dual purpose and that the straight shaft extended towards the spinner; though it is not actually visible in this picture it can be seen in a slightly later painting and in both cases could have been used for plying or twisting yarn. It is fairly certain that Lucas van Leyden produced this portrait when he was living in Leyden, which was the centre for the Dutch wool industry. From the wavy appearance of the fibres attached to the distaff one would assume they are wool and that the spinner is making a worsted-type yarn from the fibres prepared parallel and tied round the free standing distaff with a simple broad band. An almost identical spinning wheel, with a single driving band and no tensioner, is depicted in an engraving by the sixteenth-century Flemish artist Jan van Galle. Again the stock slopes from the wheel down to the spindle (reminiscent of the spindle wheel perhaps) but in this instance the orifice of the flyer is, as one would expect, facing towards the spinner and she sits turning the wheel with her right hand by a little handle on the axle, and drafting the fibres with her left, unlike the lady portrayed by Lucas van Leyden. Van Galle has treated the subject in an altogether more romantic way: entitled 'Arachne', the spinner wears flowing garments and head-dress and is surrounded by various tools used in the making of woollen cloth. As well as a loom, there is a frame holding teasles for nap-raising and a pair of croppers, while beside her in a basket is a pair of carders. She uses a free-standing distaff which has, at the top, a disc of vertical prongs so that as she draws the wool fibres the short ones get left behind. Half-way down the post, on an attached small platform, two bobbins are placed side by side on vertical spikes, for plying.

Perhaps the most famous painting of a sixteenth-century spinster is that of Anna Codde, painted in 1529 by Maerten van Heemskerck (32). At that time he was a pupil of Scorel in Haarlem, who had done much travelling in both Germany and Italy. But Heemskerck himself did not go to Italy until after this portrait was painted, so we can assume that it is a true Dutch spinning wheel of the period. It shows clearly the spindle extending beyond its supports in both directions, the flyer facing the spinner in this instance, but it also has certain characteristics not seen on the other two spinning wheels from the Low Countries so far mentioned. The spindle is kept in place by two wedges, one in each of the maidens, which allows for both easy removal and tensioning. There are two driving bands, one linked to the spindle whorl inside the front maiden, and the other to the bobbin

32 Anna Codde, 1529. After a painting by
Maerten van Heemskerck (1498–1574) in the
Rijksmuseum, Amsterdam

whorl on the outside. There is no suggestion that the bands cross; since the
details of this painting are so clear it must be assumed that they are two separate
bands, which could be tensioned independently if the spindle were wedged
more tightly into one maiden than the other. The distaff is attached to the stock
which in this case slopes down to the wheel. Haarlem, although later famous for
linens, at this period had a wool industry. Like the other ladies in the Low
Countries using the small hand-turned flyer wheel, Anna Codde appears to be
spinning long wool into a worsted-type yarn. Since the long fibres of wool shown
in the painting, with little crimp, are liable to have less elasticity than short
springy wool, the technique of spinning this type of wool can be very similar
to that of spinning flax, even though the former needs to be greasy and the
latter damp. No picture could show more clearly the spinning of long fibres
with one hand. They are drafted by the index finger and thumb, and a tension is
kept on the thread between the third finger and the orifice. One can see how the
thumb is perfectly poised to roll along the index finger, and how the wrist is
turned as the thread is smoothed before a few more fibres are selected.

Common to all these small spinning wheels is the single hoop rim of the
driving wheel adopted from the spindle wheel; only one of those mentioned has
elaborately turned spokes. With their compact size, the spinning wheels were

used in a sitting position. The spinners portrayed in these paintings are all ladies in fine clothes and therefore it seems that it was an accepted tool in all classes of society and could, although in many parts did not, replace the spindle and distaff for spinning warp yarn. Such spinning wheels must have been used in the Low Countries earlier than the dates of the pictures, since Jozef Weyns (1974) found that in an inventory list from Ghent dated 1514 there is an entry, 'a little wheel to spin with', and in one of 1519 from Courtrai, 'small wheels to spin with'.

It would seem that there were two types of mechanism on these small spinning wheels: either the bobbin lead with a double band and the flyer between the maidens, or a bobbin drag with a friction band and a single driving band to the spindle between the maidens and the flyer outside them. To what extent these represent two separate sources of evolution is difficult to assess, particularly as Anna Codde's spinning wheel contains a mixture of the two. It does, however, look as though they may have arrived in the Low Countries from two different directions, from Germany in the east and France in the west, although the second arrangement may even have been an evolution which occurred in Flanders itself.

While flax appears to be the fibre that was spun on flyer wheels in Germany and wool in the Low Countries, there is no reason to believe that flax was not also spun on them there. When flyer spinning wheels came to England it is usually said that they did so with the Flemings, yet by and large, they were not described in terms of their size but were differentiated from the spindle wheel by the type of fibre they were used for. In *Household and Farm Inventories in Oxfordshire Between 1550–1590* (edited M. A. Havinden), about one inventory in five lists spinning wheels, and these are of two different types, the wool wheel and the linen wheel, the earliest mention of the latter being 1557. A widow and gentlewoman from Moreton (in Marsh) who died in 1585 owned 'two woollen wheels, 2 linnen wheeles and a little fine linnen wheel with frame for fringe', while two years earlier a labourer from Wootton is recorded as owning 'one woollen wheele, one linen wheele, and one paire of cardes'. Thus in England too, all strata of society were using 'linen' spinning wheels before the middle of the sixteenth century.

In 1566 Anne Cecil, when aged eleven, received a New Year's gift from her father, later to be Lord Burghley. He wrote this poem to accompany the present:

> As yeres do growe so cares encreasse
> and thyne will move to loke to thrifte,
> Thogh yeres in me woorke nothing lesse
> Yet for yore yeres, and new yeres gifte
> This huswifes toy is now my shifte
> To set you on woorke some thrifte to feele
> I sende you now a spynneng wheele
>
> But oon thing firste, I wishe & pray

Leste thurste of thryfte might soone you trye
only to spynne oon pounde a daye
and play the reste, as tyme require,
Swete not fling rocke in fyre
God sende who sendeth all thrifte & welth
you long yeres & yore father helth.

There can be little doubt that the 'housewife's toy' was a small wheel, therefore presumably with a flyer and a continuous wind-on since Anne's father considered that his daughter would find it a much faster way of spinning and that it should replace her spindle and distaff. One wonders whether the word 'play' in this context means ply, although it could equally well indicate that the time the new spinning wheel would save should give her more leisure.

Before the discovery of the picture in the *Mittelalterliches Hausbuch* the invention of the flyer spinning wheel was usually attributed to Meister Jurgen, a citizen of Brunswick in Lower Saxony. It would be interesting to know why so much is made of Jurgen, and it may be simply that historians searching for an inventor seized upon a mention of this man with references connected with the spinning wheel. Whatever the reason, the expression Saxony (or Saxon) wheel, as a description for a flyer wheel, irrespective of its type of drive, shape or whether with treadle, has been widely adopted even though certain writers have ascribed one or other of these features as particular to the Saxony wheel.

According to the Chronicle of Brunswick-Lüneberg by P. J. Rehtmaier in 1722, 'Also this year 1530 the spinning wheel which the women now use, is thought of or brought here by a citizen and artist stone-carver and wood-carver with the name Meister Jurgen; which Meister, he lived in a tavern the other side of Oelper, now the tavern is still named "To the spinning wheel"'. Among the historians who have tried to find further evidence of the spinning wheel inventor through the study of Jurgen, is W. F. Schweizer who, in Holland in 1965, wrote the results of his research in an article in *Textielhistorisch Bijdragen*. He found that a similar statement to Rehtmaier's had been hand-written into the margin of the register of the Virgin Church of Brunswick by the priest at least 50 years earlier, but still over 100 years after Jurgen's death. Although Schweizer was unable to discover in what particular way Jurgen was connected with the spinning wheel, there are certain facts which are worth noting. The place beyond Oelper where Jurgen lived is the village of Wätenbüttel on the banks of the River Oker on the road north from Brunswick to Celle; not to be confused with Wolfenbüttel, also on the Oker river but south of Brunswick. The inn is still called 'Zum Spinnrad' and on a large shed opposite, Schweizer noticed a weather-cock in the shape of a little horizontal spinning wheel and dated 1521. As the shed and inn are on the edge of the village, he suggests that it may have been an advertisement that spinning wheels were made there, rather than that spinning took place, which indeed it probably did too.

However, an entry dated 1563 in the 'Verfestungsbuch Gemeiner Stadt 1525–1585' referred to a Meister Jurgen Spinnrad, thus establishing that he had the nickname 'Spinning wheel' in his lifetime. For saying unchristianlike and dreadful things he was put in a cellar and there took with him a spinning wheel, which the Town Council thought must also be 'unchristian' and so forbade people from having anything to do with it, else they too would be considered bad. Jurgen's job was making stone funeral epitaphs, a number of which can still be seen in churches within easy distance of Brunswick, but it is possible that besides this he may have had a workshop in Wätenbüttel in which he made spinning wheels, perhaps as a side-line when he was in residence. But it is impossible even to speculate over which, if any, of the many modifications of that time he may have thought of. It could have been in his very character that the myth of inventor was created; a man who lived in an inn must have enjoyed the gregarious bar life, and having picked up a few spinning wheels on his travels, he took credit to himself.

Jurgen's connection with the spinning wheel remains unclarified, but if we believe that the flyer and differential speeds were initially inspired by the silk twisting mills, then as far as Europe is concerned we must consider that the flyer spinning wheel, or at least one type of it, travelled northwards from Italy.

For the introduction of the treadle there still seems to be no definite evidence prior to the seventeenth century, from which period several pictures show spinners using two hands when working at their spinning wheels; even though the treadle itself is not always visible, its presence can be assumed.

Since Jurgen was not the inventor of the spinning wheel it is suggested that he was the man who added the treadle. Born quotes the miniature from the Glockendon Bible as evidence that Jurgen could already have known about a treadle, but Born cannot have seen the miniature; the lady has both feet resting on a crossbar under the bench and there is no hint of a connecting rod or attachment to the wheel which would indicate a treadle drive.

It is just possible that the treadle came with the vertical type of spinning wheel and was then added to the horizontal one. Looking at horizontal spinning wheels, a design of treadle can now and then be seen – for example on some Scandinavian wheels – made as a separate unit, so that the two front legs from the stock rest in recesses on the treadle bar. This is not usually found on vertical spinning wheels and when it is, it has the triangular shape associated with the horizontal types. Nor are there vertical types without treadles, save where there are signs that they have been adapted, usually into winders.

The earliest actual surviving spinning wheel with treadle may be one which is illustrated in a little booklet *Wolle spinnen am Handspinnrad* by J. Glemnitz; the wheel, dated 1604, is of one of the vertical types and is in a private collection of antiques in Germany.

Feldhaus, writing in 1914, says that it might have been in England that the treadle was invented, added to a horizontal type with sloping stock, and although he gives no reason for this statement, in *Spinnradtypen*, written in 1895 by van Rettich, such a spinning wheel is illustrated and entitled 'Old English Treadle Wheel'. It is referenced to *A Treatise on Spinning Machinery* by Andrew Gray written in Edinburgh in 1819, which shows exactly the same drawings as those in Rettich but entitled the 'Common Spinning Wheel' and, as one would expect, it resembles an early nineteenth-century Scottish wheel. It may therefore be appropriate to quote Gray's introductory remarks to this Common spinning wheel: 'The spindle and the distaff, no doubt, had been made use of in spinning long before the spinning-wheel was invented. However, as we can say nothing respecting the origin of the spindle and distaff, neither can we say anything certain in regard to the origin of the spinning-wheel, by whom it was invented, or into what country this useful machine was first introduced. Probably it has undergone several alterations since first invented, and is now brought to a considerable degree of perfection; nevertheless it may possibly still admit of some small improvement.'

Chapter Four

Accessories to the spinning wheel

THE DISTAFF

The distaff is the holder for the cleaned and prepared fibres. In many languages it is called a rock, a word that was also used in England and Scotland. Developed from a cleft stick, a branch of a tree or a straight pole ('diz' meaning decked-out), distaffs are normally made of wood, sometimes cane, in a wide variety of shapes and sizes, some elaborately carved, others simply a functional stick. When used with a spindle all types of fibres are attached to the distaff, often first pulled into a roving, including wool in particular, and then wound on to it. (In some communities where a distaff is not used, the roving is wound round the wrist; small distaffs that hang on the wrist are also known.) The distaff was often a present lovingly made and given by a man to a woman as an engagement or wedding gift, or by a father to his daughter. In parts of Germany distaffs were decorated with bells and tied to the waggon which carried the bride's trousseau in the wedding procession to add an extra touch of gaiety.

In the illustrations of spindle wheels in use, it will have been noticed that no distaff is present. This is not to say that the distaff never has been used with these wheels, particularly if spinning from a roving. However, in the Middle Ages, when distinguishing between yarn spun on the spindle wheel and that spun on the spindle, the word spindle was not used in either case, but the work was referred to as spinning on the wheel or on the distaff. This has unfortunately led a few, unfamiliar with the processes of yarn making, to believe that in the latter case the distaff did the actual twisting. *Le Livre des Mestiers* of Bruges, c.1340 includes this little rhyme written in Picard French and in Flemish (the former being given below with a literal translation):

Cecil le Fileresse	Cecil the spinner
Fu chi avore lay	was here with him,
Et elle prisa moult rofile	and she much values her thread
Qui fu filé à le Kenouille;	which was spun on the distaff;
Mais le fil	but the thread
Que on fila au rouwet	which one spins with the wheel
A trop de nues	has too many lumps [slubs]
Et elle dist qu'elle waingne	and she says that she earns
Pluis à filer estain	more to spin warp [the chain]
A le Kenouille, queà filer	at the distaff than to spin
Traime au rouwet.	weft with the wheel.

With the introduction of the flyer spinning wheel it became necessary to attach the long fibres to a distaff and it is obvious that, in many places, the same distaff was used in the same way as it was with the spindle, and continued to be so even when used with treadle wheels. This is particularly noticeable in illustrations from northern Italy and France; for instance, a late nineteenth-century photograph in the Musée Arlatan in Arles, Provence, shows a woman using a spinning wheel with the distaff, a long pole held between her knees, in exactly the same manner as a picture in the same museum of an old lady spinning with a spindle (or for that matter as portrayed by Peter Bruegel in the sixteenth century).

In the Pyrenees and Slovakia, where there was a tradition to attach the distaff to the leg of a chair – in the former on the right-hand side and in the latter on the left – it could be used both with a spindle and with a spinning wheel.

The free-standing distaffs so often found in southern Germany, Austria and Switzerland, long turned poles mounted on little stools, were also suitable for spinning with either spindle or wheel. (These distaffs usually came apart in two places, making them easy to store out of the way when not required.) We see such a distaff in use in the Low Countries with the early type of flyer spinning wheel (see 31, chapter III) but in the painting of Anna Codde (32, chapter III) the distaff is attached to the stock. The distaff thus has become an integral part of the spinning wheel, often adding elegance and balance to the overall appearance. However these distaffs do not seem to have possessed the same personal significance as those used with the spindle. (In Norway the word changes: *rokk* when used with a spindle and *rokkehode* when used with a spinning wheel.)

Although distaffs attached to spinning wheels were used for holding combed wool, by and large it is with flax that they are particularly associated. Some form of lozenge-shape is frequently found at the top of a distaff round which the fibres are wound or arranged. Diderot's *Encyclopédie* shows that in France this could be of solid wood carved in one with the stick; also illustrated is a distaff with a half-moon shape surmounting the stick over which the fibres could be placed (47). However in Britain the distaff was more often made with a shaped cage of five or six fine willow or cane sticks or brass wires which were threaded through a wooden disc to keep them in shape (33). They usually bulged out at the bottom giving a pear-shaped appearance. Variations on this theme, but not necessarily with the disc, can be seen on North American spinning wheels, while in the more southerly parts of Europe the lozenge shape is made from flat slats of cane bent round the top of the pole.

An alternative form found in Britain and the Low Countries is termed a lantern distaff and resembles an upside-down flower pot. Sometimes shaped from a solid piece of wood, it is more familiar when made with a circle of narrow wooden slats attached to two discs, the one at the top having a smaller diameter than the one at the bottom.

In Germany, Austria, also Czechoslovakia and elsewhere in Central Europe, the elaborately turned pole distaffs were also attached to spinning wheels. The

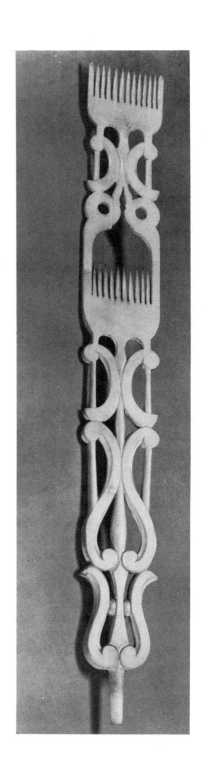

33 Scottish vertical spinning wheel. Wheel
diameter 13in. (33cm.). Pear-shaped 'cage'
distaff on the left; water pot on the right.
*Photograph: National Museum of Antiquities
of Scotland, Edinburgh*

34 Carved distaff with combs for short or long
flax, Lapland, 1879. Overall height 25in.
(63·5cm.).
*Photograph: Crown Copyright, Victoria and
Albert Museum, London*

knobs and grooves made it possible to attach the fibres at any place on the pole suitable for their length and the convenience of the spinner. The distaffs in the Lower Saxony area of northern Germany have a turned knob at the top for attaching the flax and the straight pole, which measures overall only about 24in. (61cm.), flares into a bell shape approximately 18in. (45·5cm.) lower down. This is also found in Holland. An even shorter Dutch distaff, only about 12in. (30·5cm.) long, ending in a point and with a disc fixed to it half way up, appears to have been used in the nineteenth century for holding factory-prepared rovings from which the spinner made the thread.

Plain truncheon-shaped distaffs with simply a knob at the top are also found in the Low Countries, as well as in Scotland, Norway, and in Denmark where cones of stiffened felt or some non-slippery material to hold the fibres are sometimes placed over the tops of such distaffs.

In Sweden highly decorative carved distaffs are used which fit into the arm on the spinning wheel. They are very beautiful tools, and shapes and designs vary in the different parts of the country. Every trick of the expert wood-worker is used, including turned captive balls inside little cages. In the north-east particularly, which has been the flax-growing area since the twelfth century, many of the distaffs have prongs at the top, a comb effect, to hold the flax straight down; a second comb is sometimes carved out further down which is used alternatively when the flax is shorter (34). The comb acts like a final hackle so that as the flax is drawn down any short ends remain in the prongs. Comb distaffs are also used with spinning wheels in Norway, Finland, Russia and much of Eastern Europe, and I have seen an example as far south as Switzerland.

In Russia there is a wealth of ornate carving on the distaffs. Some of these are L-shaped, having a smooth horizontal piece on which the spinner sits to keep it steady, while the vertical post is carved and holds the fibres for both spindle and wheel spinning. These are also used in Slovakia. In both Finland and Russia the L-shaped distaff sometimes has a comb at the top, and a picture in *Russische Volkskunde* (D. Zelenin) shows women from the Ukraine using vertical spinning wheels and sitting on the distaff placed on an ordinary table. Also from Finland and the Baltic lands come flat bat-like distaffs which form part of L-shaped types or are attached to sticks. From the excavations at Novgorod we know that such distaffs were in use in Russia as early as the eleventh century (Kolchin, 1968).

Distaffs for spinning tow have four or five wooden spikes fitted into a disc at the top of a short turned post. They are found particularly in Austria, Switzerland, Germany, northern Italy and Scandinavia (35). The short fibres of the tow are placed in a mass between the spikes. (In some places this distaff was also used for wool which had been first teased into a fluffy mass.)

In the nineteenth century attempts were made to improve distaffs for flax spinning. A master weaver at an establishment at Christianhavn, Copenhagen, constructed a box, its base lined with hackle teeth to hold the flax lengthwise. Movable brushes were put on top of the flax to keep the fibres parallel and to hold

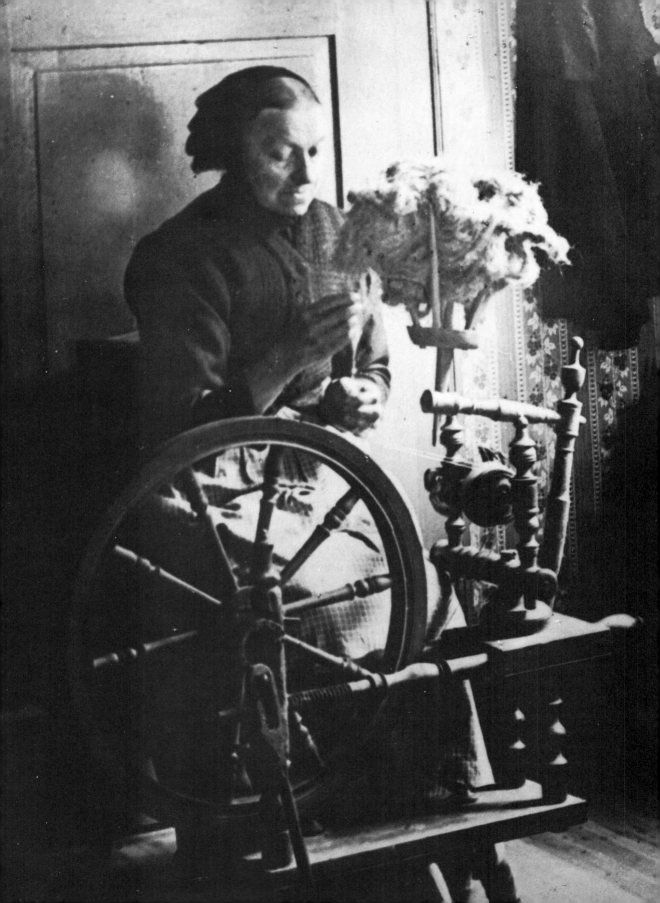

them back as a few at a time were drawn out. The box, sloping downwards towards the orifice, was supported by three posts above a vertical spinning wheel, making the whole thing look uncomfortably top heavy. It may not have been used only in Denmark (Erna Lorenzen, 1969). Perhaps the most remarkable distaff (illustrated by Rettich, 1895), if it could be so called, was invented by a citizen of Nantes in France in 1832 and won a silver medal in Paris in 1833. The flax was held lengthwise, by some grips, on a sloping board which had, at the bottom, a brass tube leading to a small kettle. This emitted steam in order to save the spinner the trouble (which Rettich describes as exhausting) of wetting her fingers with saliva.*

DRESSING THE DISTAFF

There are many methods of attaching the flax to the distaff (dressing the distaff). This partly accounts for the variety in distaffs though the arrangement of the fibres themselves follows basically one of two ways; either they are spun parallel, as they have been combed in the hackles, or they are criss-crossed over each other. The foremost countries in linen production for many centuries were Holland, Flanders and the Cambrai district of northern France, but we see from the writings of Louis Crommelin that they had different methods of dressing their distaffs. Crommelin, a Huguenot, came from a family which had been connected with the linen trade for many generations in Armandcourt, Picardy. After the Edict of Nantes in 1685, when so many cloth workers had to flee from France, Crommelin went first to Amsterdam, but in 1705 William of Orange sent him to be Overseer of the Royal Linen Manufacture in Ireland, where he settled in Lisburn with 70 other French and Dutch linen craftsmen. He there did much to improve the standards of the linen industry but he did not approve of local methods of flax preparation and spinning which, by the time he reached Ireland, were firmly established in the Dutch fashion. Horner describes this method: 'The spinstress, when about to prepare her flax for the distaff, tucks one end in the bib of her apron and taking the other end carefully by the corner lays it over in broad folds somewhat broader than the circumference of the distaff one upon the other.' Crommelin objected to this method: 'They that would spin flax ought in the first place to suit their flax to the yarn which they would spin. The people of Flanders, who spin far finer than those in Holland, do always (when they spin) tie their flax or hemp on the rock, straight down in a line, only at the lower end of their flax they give it a loose turn up and so draw the thread straight down, whereby they have the whole length of the flax. They of Holland draw out their

*In the *Statistical Accounts of Scotland*, an entry for Kirkintilloch in the County of Dunbarton reads: 'The waste of saliva in wetting the thread must deprive the stomach of a substance essential to its operations, whence all the fatal consequences of crudities, and indigestion may be expected.' In the same section, spinning, it says, is partly to blame for the females in the neighbourhood being subject to hysterics. Also, that natural health is affected by spinning since one side of the body is exposed to the chill of the season while the other is relaxed by the warmth of the fire, and as only one foot and hand are used, it causes uneven division of labour for the body.

flax in a flat cake (as it were), and then tie it on the rock, whereby their flax is spun athwart (as for your short flax it must always be spun athwart, but for your long flax, at length). The people of this country, in imitation of the latter, do all spin in that manner, both their long and short flax, which is extremely preposterous, for first, they can never spin so fine and so even a thread by taking the whole length of it; secondly they can never dispatch half the work that way, as they would do if they spun the whole length. Thirdly, their yarn when it comes to be made up into cloth, by their present way of spinning, must cotton [i.e. acquire a raised surface] much more than if it was spun in our manner, because by spinning across the flax there are more ends everywhere appearing in the cloth which rise and start up therein, than there would be if you drew your thread down the whole length of the flax.'

To produce the very fine thread so much sought after by Crommelin may not have appealed to the women of Ireland, who were perhaps not prepared to suffer either the discomfort or the laborious methods of spinning flax so fine. Horner mentions that in Flanders flax was spun in damp cellars and Mrs Palliser (1875) paints a grim picture of the conditions of spinners making the hair-fine thread required for lace: 'The finest quality is spun in dark underground rooms for contact with the dry air causes the thread to break. So fine is it as almost to escape the sight. The feel of the thread as it passes through the fingers is the surest guide. The thread-spinner closely examines every inch from her distaff and when an unequality occurs, stops her wheel to repair the mischief. Every artificial help is given to the eye. A background of dark paper is placed to throw out the thread and the room so arranged as to admit one single ray of light upon the work.' (In 1790 thread for lace-making in England was imported from Antwerp for as much as £70 a pound – Mincoff, 1907.)

In Ireland when Horner asked old people why they continued to dress the distaff in the Dutch manner, the only reasonable reply he obtained was that the fibres were held together in a mass, and therefore the spinner had more control over them and they did not come down in too large quantities. However Horner continues: 'No difficulty seems to be experienced by the spinners of continental countries; the flax does not deliver too rapidly; even in the spinning of the exceedingly fine yarns. . . .' The tall continental distaffs were designed and are ideal for the straight type of dressing, but in Lower Saxony where, as we have already noted, the distaffs are shorter, fibres are arranged across one another in what appears to have been the Dutch fashion. Both English and French descriptions of dressing the distaff talk of layers of flax spread out on a table and the distaff rolled diagonally across the web of fibres (forming a cone shape), but the Encyclopédie states that only long flax should be dressed in this manner and that short flax should be straight. This is quite the opposite view to Crommelin's, but for ordinary purposes it is a fact that a fine thread can be spun from either of these arrangements if the spinner is skilled. For distaffs with combs at the top it is sometimes the custom of the people that use them to take a finger of flax (about 2oz., 58g.) and to fold it up concertina-style; it can then be tied and left. When it is shaken out the flax is creased into kinks, so there is an overlapping of fibres at intervals down their length, one end of which is attached to the comb.

A rather different method was used for the bat-like distaffs and is described in some detail by Veera Vallinheimo (1956). In eastern Finland the flax was pulled out into small bunches which were then layered backwards and forwards on a table, forming a long oblong carefully smoothed either side. It was then rolled up loosely, like a long rolag, and tied to the distaff. If pulled from the bottom of the roll the fibres are drafted at right-angles to their length.

RIBBONS AND BANDS

Once the flax is arranged on the distaff, to hold it in place a piece of ribbon, approximately $\frac{3}{4}$–1in. wide (2–2·5cm.) and two yards in length (183cm.), is tied round the top and spiralled several times around the bundle of flax and the ends tied at the bottom. (Traditionally the colour of the ribbon denoted a married or unmarried woman; green or blue for the former and red, pink or white for the latter.) This method of tying the flax was used in many European countries, particularly the British Isles and France.

In Finland and countries where a bat-like distaff was used, the rolled flax was secured to it with a fine ribbon threaded in and out of the three holes drilled in the bat as shown in figure 7. (Sometimes there are only two holes, in which case the ribbon goes right round the bat.)

An alternative method of holding the fibres in place, prevalent in the Low Countries, Germany, Switzerland and Scandinavia, was by a single broad band of stiffened paper, parchment or linen, about 22in. (56cm.) long and 6–7in. (15–18cm.) wide (as illustrated in use by Anna Codde **32**, page 89). It was wound round the middle of the bundle and a tape or ribbon attached to one end was also wound round and pinned in place. Since it was seldom or never used in England

Figure 7 Roll of flax tied to a bat distaff

we have no name for it here, but in Germany such a band is called a *Kunkel* or *Kundelband, Rocken* or *Wokenbrief*. It was painted in gay colours with all sorts of decorative patterns which often included mottos such as 'Laughter makes you happy, laughter makes you rich' or 'If you do not want to spin you will not have linen'; also many allusions to the perfect and pious housewife. These *Kunkelbänder*, used instead of ribbons, were often given by the boys to their girlfriends. The fastening pins, made of ivory or wood and often decoratively carved, were tied to the end of the tape or ribbon; plainer pins were of metal.

In the areas round Strasbourg and Nancy, in France but much influenced by Germany, a long decorative band about 3in. (7·6cm.) wide was used, tied at one end to the extreme top of a tall distaff, then spiralled once round and tied firmly round at the bottom. In Switzerland a similar wide band was sometimes used and simply spiralled round the flax, and yet another method in Germany used a short, narrow band round the top and a second round the bottom of the cone.

In Sweden the flax was sometimes held in place by binding it round with the skin of an eel, which was tied in place with ribbon. Richard Hall in 1724 noticed: 'The spinners in Holland have a piece of fine woollen cloth pasted round the Rock, purposely to prevent the flax from coming down too fast, as the spinner draws; besides they have a piece of Oyl-cloth which they bind gently over the flax, while it is tied to the rock to prevent its drying too fast, and to preserve it from dust'. This under-cloth can be seen quite clearly in the engraving of a spinner by Geertruyt Rogman and there is a small cloth too, tied round the top of the flax, which may have been kept damp (36). Here the flax appears to be folded over the top of the distaff, a method, perhaps, of dressing a fairly short

distaff with long flax. Hincks' engraving of spinners in eighteenth-century Northern Ireland (37) shows that there a piece of cloth, no doubt also dampened, was placed over the top part of the flax. While being spun, flax produces a fine dust which can become unhealthy for the spinner. The cloth, while keeping the fibres damp, may well have helped to prevent the dust from flying about, as much as keeping the fibres clean.

36 The Spinner. After an engraving by Geertruyt Rogman. Mid-seventeenth century

37 After Plate VI of a series of 12 engravings illustrating the processes in the Northern Ireland linen industry, 1783 by William Hincks. It shows spinning, reeling with a clock reel and boiling the yarn. Co. Down

WATER POTS

Flax spun dry makes a thread which is rough, with ends of fibres sticking out giving a bristly effect. Dampening with saliva was done either by passing the thread through the mouth or wetting the thumb with the tongue or lower lip. While this was quite acceptable amongst a peasant society, it was not suitable for ladies spinning in their parlours or drawing-rooms, so little pots filled with water were placed convenient to the spinner's hand. In Germany they were often made of pewter in a circular shape to fit on to the lower part of the distaff, forming a little trough all the way round; or it would be carved out of the wood itself in much the same shape. In Courbet's delightful painting of the 'Sleeping Spinner', a little round pewter pot hangs just beside the orifice, so that the girl can dip her fingers into the water before she slides them up the formed thread to make it smooth. It is interesting that there should be these different positions for the water, as they must influence the rhythm of the spinning. In Belgium and

Italy there are neat little water pots on a curved handle, like miniature buckets, which hang on the mother-of-all just beneath the orifice.

Sometimes a recess on the stock was provided to hold a little dish for water or an egg-cup-shaped receptacle was attached to the stock or held by a separate holder, such as an articulated arm, also attached to some part of the spinning wheel, usually on the opposite side to the distaff (33).*

GREASE POTS

It is thought that some of the small receptacles found attached to spinning wheels were not for holding water but were filled with grease for lubricating the wheel axle, spindle and leather bearings, a vital part of caring for the tool. One also meets separate grease pots; one from Scotland was found hanging from an upright on a wheel; the pot, made of horn with a wooden plug at the bottom, was similar to those used by sailmakers for greasing their needles, to help them through tough canvas. Such grease pots were also used in the Isle of Man and were kept, it is said, in the centre of the stock where there are little recesses cut to different shapes, diamond, round, quatrefoil and square.

SEATS

A most important accessory for the flyer spinning wheel, whether it be turned by hand or treadle, is some form of seat. Most pictures show spinners using an ordinary wooden chair. Stools, especially with hand-grips, were obviously much used; as spinning was so often done at the cottage door for the better light, a small spinning wheel, particularly the vertical type, could be carried out in one hand and the stool gripped in the other. In England and Scotland a stool-like chair with three legs, sometimes four, and a narrow straight back was used for spinning. Sometimes these were very ornately carved, but Victorian copies, often quite plain, can still be found in antique shops. The heights of the seats vary from 13in. (33cm.) to 18in. (45·5cm.) according to the height off the ground of the orifice of the spinning wheels they were intended to be used with.

Lower Saxony is an area where there is a special chair for spinning, as also is the Tyrol (38). In the Landesmuseum of Brunswick there are six examples, one dating from as early as 1677, the others from the eighteenth and nineteenth centuries. Their feature is the lack of an arm on the left-hand side of the chair, thus allowing the spinner freedom of movement with the hand and arm which does most of the work. The arm and hand movements of flax spinning are not energetic ones and it must have been comfortable to sit spinning beside the fire in one of these chairs with an arm rest on the right. The seats are made of rush. They measure about 19in. (48cm.) across the front and vary from 15–18in. (38–45·5cm.) off the ground.

*In the drawer, too small to hold bobbins, of a late eighteenth-century spinning wheel there was a tiny sponge attached to an ivory handle. I can only suppose this was dipped in water and then brushed on to the fibres to dampen them. The *Encyclopédie* mentions a *mouillette*, a sponge moistened with water kept within reach in a little porcelain or tin vase.

38 Spinning chair from the Tyrol, 1841.
*Photograph: Tiroler Volkskunst-museum,
Innsbruck*

BOBBIN HOLDERS

Once the bobbin is full of yarn it has either to be emptied or set aside to be used in conjunction with other bobbin-fulls for plying. For this purpose the full bobbins are placed on a rack beside the spinner. In England this rack is called a Lazy Kate (*figure* 8) and in Scotland a whirrie. The bobbins can either lie alongside or above one another according to the design of the Lazy Kate. In Holland it is made in the shape of a box which has a slanting back, where the bobbins are held one above each other but slightly staggered. Some spinning wheels are provided with horizontal bars between uprights or vertical spikes on cross-bars to store spare or full bobbins and it would be possible to ply from them.

A feature of some Dutch and Belgian spinning wheels is a cross-bar with two to four vertical spikes for holding the bobbins, inserted in place of the distaff for plying (*figure* 9).*

---

*In the Stranger's Hall, Norwich, there is a beautiful example of a late eighteenth-century vertical spinning wheel, painted dark green and decorated with Chinese figures in gold lacquer. Two bobbins, placed one above the other on the left, and a little above the orifice, replace the top of the distaff. One feels this wheel might have been used for silk twisting.

Figure 8 *Lazy Kake (after a nineteenth-century example from Sweden)*

Figure 9 *A distaff from the Low Countries for holding bobbins for plying*

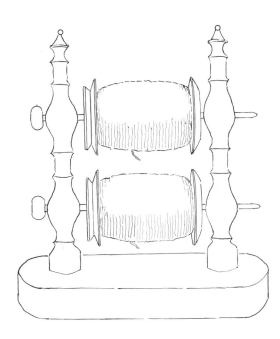

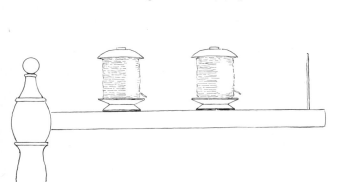

REELS

The simplest form of reel for taking the yarn off the bobbin was a straight stick with two cross-bars, one at either end, around which the yarn was wound (*figure* 10). However, if one of the cross-bars is put at right angles to the other it forms a cross reel, or what is commonly called a niddy-noddy (see **36**, an example beside the spinner's chair). To wind yarn on to this, it is held by one hand in the middle of the central stick and rocked backwards and forwards – hence its name – while the yarn is guided by the other hand. Sometimes the central stick is extended at one end to form a handle by which it is turned. With either arrangement the yarn is wound in a circular motion, going over and under the cross-bars alternately, so that the hank when removed is four times the length of the stick. If the distance between the two cross-bars is known, the yardage of the hank can easily be calculated, but this is not very accurate on account of the build-up of yarn on the cross-bars. Cross reels can be either very simply made, or smoothly turned, sometimes decoratively carved, while the cross-bars may be curved in such a manner that the yarn stays better in position. One end is straight for slipping the hank off without stretching it.

To avoid having the spinning wheel idle, it was customary to remove the full bobbin and hand it over to someone else to reel. In this case the bobbin was placed on a wand or pin, a spike with a handle, on which it could rotate freely. The wand was held in one hand while the reel was rocked with the other.

39 Free standing clock or wrap reel.
Circumference 90in. (230cm.).
*Photograph: Crown Copyright, Science
Museum, London*

The rotary reel is a quicker and more accurate tool than the cross reel, and therefore it became obligatory in industry. It comes in many different forms, but basically it is similar to a rimless wheel, attached to a single support with four or six spokes each with a cross-bar at the end to hold the yarn (39). It is turned by a knob fixed to one of the spokes, and the end of one spoke is usually made collapsible to permit easy removal of the hank. The central axle has a thread (worm gear) which connects with one wooden cog wheel which has a small wooden peg on its side. This passes over a spring stick which makes a loud click indicating that the reel has turned a certain number of times, and according to the circumference of the reel the yardage is known. Alternatively the cog turns the hand of a clock, but it is equally usual to have two cogs, so that there is a click and a clock to indicate how many times it has clicked! In Germany even more complicated arrangements can be found, with hammers that hit bells each time. The rotary reel, sometimes called wrap reel, click or clock reel, can be free-standing or attached to a board which can be placed either on a table or across a chair seat. (A free-standing clock reel can be seen in use in Hincks' engraving, with the bobbin held on a pin, see 37, p. 103.)

Another type of reel has two sets of four spokes on a central axle with cross-bars connecting the parallel spokes; it is turned by a handle on the axle. A French one illustrated in the *Encyclopédie* (47) sits on a table and there is provision for the pin, which holds the bobbin, on the left of the reel.

From an eighteenth-century English engraving of the woollen industry (*Universal Magazine* 1749) when the great wheel was used for spinning, it is possible to see that the yarn was built up on the spindle into a cone, here called a coppin. 'It is taken off without ravelling, and carried on another spindle to the other work-woman who winds it into skains on a reel', reads the description. A disc must have first been slipped on to the base of the spindle, since the reeler holds the cone by the disc and the yarn unravels from the point. The reel is a large version of the type described above but the woman turns it with her left hand and holds the disc with her right.

Similar to this but constructed differently was a reel used in the peasant communities of North Italy, Austria and Germany. Two rectangular frames are slid on to a horizontal post which is also the axle, with the handle at one end. The frames are at right angles to each other, thus making the four arms and uprights support the axle post at either end. Such reels vary in size from those which stand on the ground to the smaller type for placing on a table (40). To remove the hank of yarn when completed, the frames can be pushed together so that the peripheral distance becomes less and the hank can then be slipped off.

Occasionally one finds rotary reels attached to spinning wheels. The latter are usually vertical types and the reels are placed above or behind the wheel; I have seen examples from Bavaria, Austria, Italy, Norway and Holland. Horner found

Figure 10 *Stick reel (after an example from Crete)*

one attached to a horizontal spinning wheel from the region of Kempen, in Germany just across the border from Holland, but compared with the Dutch arrangement the reel looks cumbersome. The reels can of course be used simultaneously with spinning, but the bobbin must be very evenly wound so that the yarn winds off evenly on to the reel. A grooved whorl on the reel is linked by a continuous band to a grooved whorl on the wheel axle between the upright and the top of the footman.

The size of the hank is determined by the circumference of the reel, and although 90in. (230cm.) was the most usual there are many variations, as also in the number of turns between each click. After each click, that particular number of threads is tied together. This unit can be called a lea, a cut, a wrap or a knot. Another set of threads is then started beside the first, not on top of it, to avoid the build-up which will distort the measurement.

The standardisation of the reel varied from district to district, and there could even be more than one size in each. In the Norwich area, for instance, in an Act of 1791 to prevent abuses and frauds committed by persons employed in the manufacture of combing wool and worsted yarn, there was mention of a yard reel – 36in., a $7\frac{1}{4}$ reel – 63in., and an $8\frac{1}{4}$ reel – 72in. Each wrap or lea contained 80 threads and the official amounts were 7 leas on the yard reel, six on the $7\frac{1}{4}$, and

40 Reel with clock indicator, 1845, from Niederbayern, (Lower Bavaria).
Photograph: Museum für Deutsche Volkskunde, Berlin

7 on the $8\frac{1}{4}$; these had to be wrapped in pounds.

William Murphy (1910) states that of the many different tables, the oldest one known in Britain is the following for linen:

1 thread	=	1 reel turn	=	90in.
120 threads	=	1 lea or cut	=	300 yards
12 cuts	=	1 hank	=	3,600 yards
200 cuts (or 16 hanks)	=	1 bundle	=	60,000 yards
20 hanks	=	1 reel	=	72,000 yards
50 hanks	=	1 three-bundle bunch	=	180,000 yards

The tables were based on length only, but from them was developed the system of counts, an arbitrary table which told the thickness of the yarn (in other words, the yarn diameter) by the number of leas or cuts that could be spun from a pound of raw material. The higher the number, the finer the yarn.*

For example, using the above linen table, 10 lea would be 10 × 300 yards = 3,000 yards of yarn in one pound, while 20 lea would be much finer since there would be 6,000 yards in one pound. If the yarn is plyed, the ply number is put before the singles number, and accordingly the thickness of the yarn is changed, e.g. 2/10s = 5s since it is two times as thick as the singles. (The exception to this is spun silk when the number of yarns plyed is put before the resulting count number, e.g. 2/10s = 10s because it is two threads of 20s twisted together.)

An old measuring term often found in Scotland is a spyndle of yarn, which can be 14,400 yards by the Aberdeen table and 11,520 yards by the Stirling table. Richard Arkwright produced a reel, to be seen in the Science Museum, London, which was used for cotton and measures the hanks in 840 yards. (This is also the measurement for spun silk.)

300 yards remained the standard lea for linen. Other standards are 560 yards per hank for worsted; 256 yards per skein for Yorkshire wool; 320 yards per snap for West of England wool; 200 yards per cut for Galashiels wool.

Today there is an international system for all fibres called Tex, which is the weight in grams per 1,000 metres.

Once the yarn is in hank form it is ready for scouring, dyeing or whatever is necessary, and it is usually stored like this.

SWIFTS

The simplest type of hank holder for winding yarn into balls was made from the trunk of a small tree, cut a little above the point where it forms its branches, the stumps of which make a stand. To the centre of the trunk are attached two flat

*Since these are old systems all based on yardage per pound, the metric equivalent has not been given here. If requiring to calculate the count number of an evenly hand-spun yarn using one of these systems, measure the yardage and weigh the hank. To find the weight of a pound (assuming that the weight is measured in ounces) divide the yardage by the weight and multiply the result by 16. Divide this by the count number you are using.

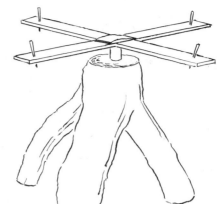

Figure 11   i. *Hank holder (after an example from the Isle of Man)*
      ii. *Swift (after an example from Cyprus)*
      iii. *Rice (after an example of an early nineteenth-century table model made of mahogany and box-wood)*

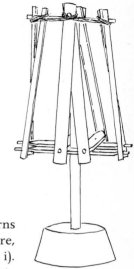

ii

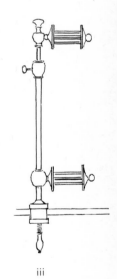

iii

lengths of wood set at right angles to each other and forming a cross which turns about its centre. Holes in each of the four arms, equidistant from the centre, each hold a little upright peg around which the hank is placed (*figure* 11 i). Usually there are three or four holes in each arm so that the pegs can be moved to accommodate different-sized hanks. More elaborate versions with folding arms were made for use in drawing-rooms and are seen in Germany and France. In England hank or skein holders, often referred to as wool winders, can take the form of a short central post with four arms which concertina out to hold different sizes of hanks. In all these types there is usually a small wooden cup on top of the central axle for holding a ball of half-wound yarn. Delightful examples of winders, often of mahogany, can still be found in antique shops.

While winders are used more by knitters and embroiderers, swifts are associated mostly with weavers and spinners. The swift is a frame of criss-cross sticks (or simply four arms) which turns on a central post (*figure* 11 ii). The frame is slightly conical so that the hanks cannot slip off the bottom, but at the same time it is able, without any adjustment, to hold hanks of various circumferences. The central post is set in a block of wood which stands on the floor. The umbrella swift, which either stands or is clamped to a table, has the wooden slats of a criss-cross frame joined in the middle and tied at the ends, so as to be flexible. When not in use this frame collapses round the central post, but when required it is opened to hold the hank by pushing a wooden collar up the central post, the collar being then locked in place by a wooden screw. On some models it is held up by a peg pushed into one of several holes bored in the central post; in Scandinavia some swifts are pushed downwards, not upwards.

Another type is the drum swift or wool rice, usually free-standing, although smaller versions are clamped to the table (*figure* 11 iii). The two drums are either solid or made like round cages. The hank goes round both and is thus held vertically. One of the drums can be adjusted to hold different sized hanks by sliding it up or down in a slot in the central post and tightening it with a plaque of wood and a screw or a screw only. The two cages or drums rotate as the yarn is wound off, and sometimes there is a double pair of drums (double wool rice) for holding two hanks of equal length. It must be added that yarn is often hanked before it is plyed, in which case a double rice is particularly useful.

Chapter Five

Principal types of spinning wheels

Within the two basic constructions of flyer spinning wheels, the horizontal and the vertical, there are certain characteristics that warrant further division. The horizontal fall into three groups: those that have a Picardy type flyer, those that are built on to a stock as described in chapter III, and those that have a frame construction. Vertical spinning wheels are built with either a base or a frame, and although Picardy type flyers are not unknown on the vertical types, they are comparatively rare and will not be discussed here.

PICARDY SPINNING WHEEL

This spinning wheel is a direct development from the early spinning wheels depicted by the Dutch and Flemish painters of the first part of the sixteenth century. Built on a stock, its essential characteristic is that the flyer and bobbin are on an extension of the spindle in front of the maidens (*figure* 12).

Two Dutch paintings, one from the late sixteenth century by I. N. van Swanenburgh (illustrated on the cover of *Ciba Review* no. 48) and the other from the seventeenth century by Gerard Dou, of Rembrandt's mother (**41**), both show spinning wheels with a larger and wider hoop rim than that of Anna Codde's and with the Picardy flyer. In the first picture the flyer is not actually visible, but there is only a whorl between the two maidens and the driving band is a single one, so the Picardy flyer can be assumed. Despite the larger rim the spinner sits to work and turns the wheel with her right hand by a knob on one of the spokes. In both paintings the distaff is made in the form of a wooden cup which holds a ball of tops prepared by the comber for worsted-type spinning. One end of the roving is bound to the distaff post to hold it tight so that the spinner can draw but a few fibres at a time to make a fine thread. This binding thus did the job of the spinner's other hand before the treadle freed it from turning the wheel, but it may also indicate that the combed wool in this case had a shorter staple length than that being spun by Anna Codde, who appears to be having no difficulty in controlling with one hand the long wool attached to her distaff.

In Velasquez's famous painting 'The Spinners', executed about 1657, towards the end of his life, it is probably a Picardy wheel that features so prominently. The position of the flyer is not entirely clear, but while the driving wheels shown

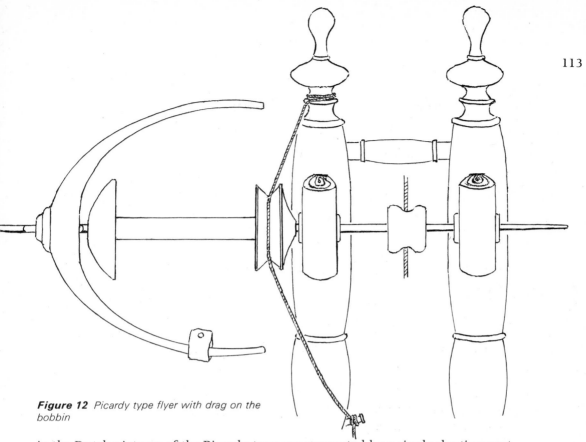

Figure 12 *Picardy type flyer with drag on the bobbin*

in the Dutch pictures of the Picardy type are supported by a single slanting post, the example depicted by Velasquez has two wheel uprights and stays on the spindle side to take the strain from this side: a feature generally associated with Picardy wheels. The painting represents a tapestry workshop, probably of Santa Isobel, Madrid (*Ciba Review* no. 20), so the spinning wheel is being used for wool. This is prepared into a roving and wound round the top of a long pole distaff which is tucked under the spinner's left arm. Tapestry weaving was set up in Spain at the beginning of the seventeenth century by Flemish weavers, so the wheel would have most probably been brought along with their other equipment or else made locally to a Flemish design.

It may be just chance that the early version of the Picardy type spinning wheel is portrayed only in the Low Countries; it could be that they were used at an equally early date in France and on account of the name could indeed have originated there. Picardy spinning wheel is the name still used when this arrangement of the flyer is present, whether in France or elsewhere.

Jean Marie Roland de la Platière, Inspecteur Général des Manufactures de Picardie, illustrates and describes a Picardy wheel in his book of 1780. He says that spinning with this type of wheel turned by hand was practically the only method used in Picardy. He points out that it differs from the great wheel used for spinning wool for *draperie*, and also from the little wheel used for flax and hemp which has many little hooks on the flyer; the hooks on the Picardy flyer are, he says, kept few to avoid the wool yarn catching.

In conjunction with the few hooks, or with a flyer with no hooks at all, de la Platière illustrates a loop of material through which the spun yarn passes, which can be slid to any place on the flyer arm to guide the yarn on to the bobbin. The flyer is, of course, positioned in front of the maidens (or *poupées*, dolls, as he calls them) and the orifice is a separate small hollow cylinder, with an eye at the side, which slips on to what must be a tapered spindle.

The wheel diameter is 23–24in. (58–61cm.) but, says Roland de la Platière, it is better if it is 24–25in. (61–63·5cm.). The bench is long and parallel to the ground. The driving band, he continues, is usually made of wool, which makes for smooth running although gut is less susceptible to atmospheric influences. For the same reason leather or straw is used for the bearings rather than hat felt, which swells with humidity. Although the methods in Saxony, Linz and Branden-burg March and Berlin are superior because they have wheels with a treadle and so both hands can be used, even so Roland de la Platière maintains that a very fine thread can be made on the Picardy wheel and that very good spinning was done in France.

The distaff is a three-pronged stick and has combed wool (12–15in., 30·5–38cm., in length according to de la Platière) folded over the prongs and looped up

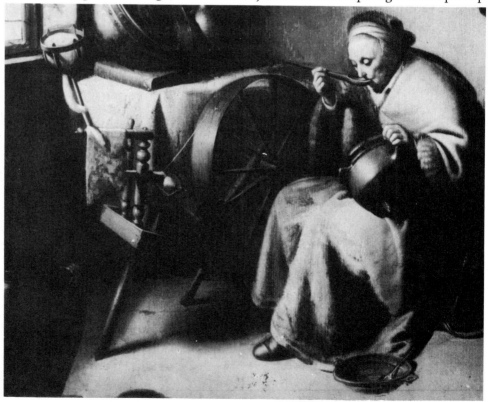

41 'Rembrandt's Mother', c.1630 by Gerard Dou (1613–1675).
Photograph: Staatliches Museum, Schwerin

and held in place by a wide band (similar to a *Kunkelband*). The distaff is tucked into the spinner's belt with the fibres on her right-hand side, so that she draws them from the bottom towards the orifice with her left hand (across her body) while with her right hand she turns the wheel by the handle on its axle.

It would seem therefore that this wheel was used for worsted-type spinning, yet it was the same wheel (turned by hand) that Louis Crommelin tried unsuccessfully to introduce into Ireland at the beginning of the eighteenth century for spinning flax. Since Crommelin came from Picardy, this indicates that despite Roland de la Platière's account, the Picardy wheel was used in that part of France also for flax. Crommelin indeed felt strongly that the Picardy wheel was the only type suitable for fine linen and thoroughly disapproved of the spinning wheel with treadle. (If the spinner was constantly stopping to examine the thread, as Mrs Palliser says, p. 100, one can understand why.)

Of the two Picardy wheels found by Horner and now in the Ulster Museum, one, said to be from the late eighteenth century, is turned by a handle, the diameter of the wheel being 23¾in. (60cm.). It has a distaff, a single turned post with a knob at the top and lozenge-shaped below, which slips into a hole on the level stock to the right of the flyer. The other, of a later period, has a treadle in addition to a handle, and no distaff. Both have stays on the spindle side of the wheel (as in the Velasquez painting), the bearings are wide pieces of folded leather, and the maidens slope away from the wheel. The orifices are extremely narrow, only ⅛in. (3mm.) and therefore it was possible to spin only a fine smooth thread. On the later example the friction band is made of leather.

An example of a Picardy wheel without treadle in the Farnham Museum, Surrey, probably from the nineteenth century, reveals one or two interesting points. The spindle 13½in. (34cm.) in length and slightly tapered, has, about 1½in. (4cm.) from the minute orifice, a short square section. This can grip the flyer with a push-on fit, yet make it easily removable for changing the bobbin. The flyer itself may not be original but it is made without hooks, and a small piece of material, with a metal eye sewn on, can be slid up and down one flyer arm. An iron nail at the bottom of the front maiden may have been inserted subsequently for attaching the friction band: but on the near front stay supporting the wheel uprights is a hole of a suitable size to hold a peg. It seems likely that the friction band went from the top of the front maiden, underneath the bobbin whorl, and was tightened on the front stay. The alignment is workable and in this manner sufficient cord would touch the whorl to effect a brake. Tensioning the driving band would be by altering the leather bearings as described on p. 56, chapter II.

Picardy wheels can be found in Canada in areas where French families settled, but there are examples known to be from the nineteenth century which have a treadle and no handle (H. B. & D. K. Burnham).

HORIZONTAL SPINNING WHEELS WITH STOCK

The flyer spinning wheel with stock, two whorls and treadle, which typifies so many people's conception of a traditional spinning wheel, is often referred to as the Dutch wheel after the country in which it was possibly developed. It was widely copied in Europe, usually being considered the older type. Crommelin also disapproved of the Dutch wheels that went with what he called 'two cords or strings', because, he maintained, they over-twisted the yarn. If these disparaging remarks were not made simply from prejudice it could have been that on the ones that he saw in Ireland the difference between the circumferences of the two whorls was insufficient to give the necessary pull-in to the yarn, and might have been more a fault of the wheelwright than of the spinner. In 1724 when Richard Hall visited Holland, he tried to discover the reasons for the success of the Dutch spinning wheels. He wrote: 'Flax wheels require to be made with an exact proportion between the wheel, the whirl, the spool and the flyer. I have endeavoured to inform myself of the true geometrical proportions of each of these, by discoursing with the ablest wheelwrights that I could meet with, wherever I came; but it has been hitherto all in vain: for either they could not, or would not tell how and in what manner this mystery in the trade, might be brought to a more mechanick regulation than at present it is under.'

Hall also noted that the Dutch wheels had firm heavy rims. In fact many of the spinning wheels still in existence in the Netherlands have almost solid driving

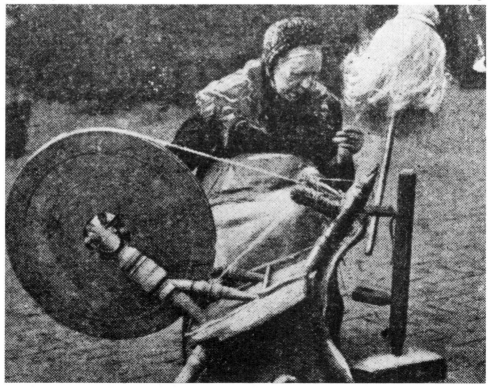

42 An old lady from the Courtrai district of Belgium spinning flax using a hand-turned disc wheel. After an old photograph

wheels. This was not confined to Holland, for in Flanders too there are similar wheels with minute spokes round the hub (42).*

It may be worth mentioning that although most of the spinning wheels seen in Belgian museums today have the doubled band drive, the spinning wheel collected by Horner from close by in Cambrai, although with a treadle, has a Picardy type flyer. The same is just discernible on the spinning wheel from Courtrai in plate 42. Both these districts have been famous for their fine linen and the drives on the wheels may bear out Crommelin's point of view.

In both Belgium and Holland there are wheels with much narrower rims and even a second inner circle necessitating a double row of spokes, giving a filigree effect. In neither case are the wheels particularly large (diameters 14–15in., 35·5–38cm.), and there is a doubled band drive. Quite characteristic of Dutch spinning wheels is a treadle with a curved bar from the left of the treadle bar to the link with the footman. Many of the existing spinning wheels come from the eastern part of Holland, Enschede having been the centre of the Dutch flax industry, but in the seventeenth century this was also an important industry in Amsterdam and there was a spinning wheel market on the Keizersgracht. Stone plaques over doorways, advertising the trade of the occupant, can still be seen in this city, dating from that period. One such plaque in St Jacobsstraat shows a horizontal spinning wheel with sloping stock and the wheel jutting out beyond it; the rim is made with felloes but is not excessively wide. The legs and distaff upright are elaborately turned, the distaff itself being bell-shaped. Another plaque, in Madeleivinstraat, announces a 'twijnwiel' or twisting wheel; beside the wheel, which has a hoop rim, are several bobbins. Although this could point to a silk throwster it is more likely to signify someone who plyed linen. Many of what we think of as spinning wheels must have been put solely to this use; the size of the orifice can sometimes be a guide – a narrow one for spinning and a wider one for plying – and in Holland there was a different type of distaff (as already noted in chapter IV).

Writing in the same century Andrew Yarraton comments: 'I say the fine Linens are made in Holland, and Flanders, that is, woven and whitened there, but the Thread that makes them comes out of Germany from Saxony, Bohemia, and other parts thereabouts. . . . Their wheels go all by the foot . . . whereby the action or motion is very easie and delightful.' The horizontal spinning wheels of Lower Saxony have much the same shape as those in Holland but on the whole a steeper slope to the stock and a wheel diameter of 15–16in. (38–40·5cm.) (43). There was a strong tradition in Germany, as elsewhere, for a girl to spin for her trousseau and a spinning wheel was amongst customary wedding gifts. Around Hanover these bride's spinning wheels (*Braut-spinnrad*) are similar in shape to

---

*Complete disc wheels with no spokes at all are found from time to time, made either from one solid piece of wood or in three pieces (occasionally more). In Norway, for instance, a home-made disc wheel in the Norsk Museum, Oslo, was used for hemp spinning (sometimes called a hemp-horse, 'hampehest'). An example from Mordaland has little oblongs and crosses cut out of the disc.

43 Spinning wheel with bell-shaped distaff
from the Hanover area of Lower Saxony.
Wheel diameter approximately 16in. (40·5cm.).
(Bobbin, spindle whorl and driving band
missing.)
*Photograph: Museum für Deutsche Volkskunde,
Berlin*

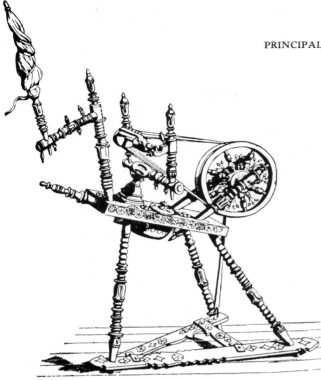

Figure 13 *Horizontal spinning wheel from Switzerland. (Courtesy of the Dicziunari Rumantsch Grischun.)*

the others but the simple lines are lost in decoration. This may take the form of little wooden birds perched along the treadle, distaff and stock, while little wooden bells dangle from any cross-bars. The wheels have decorative points between the spokes and the edge of the stock is decorated with rosettes. An example in the Historisches Museum, Hanover, painted yellow, green and red, was made in 1920; as with some Dutch wheels mentioned above, the wheel has a second smaller circle joining the tips of the points. Although many spinning wheels show the natural colour of the wood, or are stained, painted ones usually indicate a wedding or engagement present. Sometimes paint is used only to highlight the wheel points, or different colours may be traced out round the side of the rim or round the stock.

Bone and ivory studs are a favourite decoration for spinning wheels and occasionally can be found covering up blemishes. Such decorative details are not unique to Germany, however.

Further south, in Bavaria the wheels are small with diameters of 12–13in. (30·5–33cm.) and the stock is steeply sloped and often slotted for the wheel rim, making a very compact design. In south Switzerland one or two similar spinning wheels still exist, but an example from St Gallen has less slope and no slot. They have a doubled band drive and are known for some unexplained reason as English spinning wheels (*figure* 13).

From Poland and Russia Horner brought back horizontal spinning wheels with sloping stocks and doubled band drive, showing only little variation from those of Holland and Germany.

In Scandinavia there are preserved horizontal spinning wheels with a driving wheel of approximately 15–17in. (38–43cm.). Though without a markedly wide rim (about 3½in., 9cm.) they otherwise resemble those of Holland (such as on the plaque described above) and they are thought to represent the earliest type. But the spinning wheels which we chiefly associate with Scandinavia today differ from the Dutch – and from most others in Europe – in having, as their main feature, wheels that are much larger, reaching 25–26in. (63·5–66cm.) in diameter. These were developed between the late eighteenth century and the first part of the nineteenth century. The larger wheel reduces the amount of treadling and is particularly suitable for woollen spinning, though these spinning wheels are of course also used for flax and worsted. To support the wheel uprights, stays, though often found elsewhere on quite small wheels, become necessary with the enlarged wheel if the stock is to remain at a sloping angle. A Norwegian example with sloping stock has two pairs of stays (44). One pair supports the wheel uprights directly from the right-hand legs. The other pair extends horizontally to the stock, to which it is attached close to the mother-of-all; the front one of this pair is used for holding spare bobbins.

A familiar shape of spinning wheel from Norway appears to date from the first quarter of the nineteenth century. The stock is parallel with the ground. It usually measures about 20 × 6in. (51 × 15cm.) but in some examples is wider, with the front gracefully curved to allow room for the spinner's leg when treadling. The wheel is supported by two absolutely vertical uprights. On the spinner's left, four short vertical posts support a small second stock which holds the flyer mechanism and tensioner. Two horizontal wooden rods lead from this to the wheel uprights, one on either side of the wheel, and one of them, some-times both, has a screw-thread into the wheel upright whereby adjustment to the alignment of the wheel can be made by turning the rod. This particular type is also used in Sweden and in Denmark (see plate 35 on page 98) which for so long dominated Norway.

Another Danish type, believed to date from the middle of the nineteenth century, has a steeply sloping stock which is slotted to take the lower circum-ference of the wheel, a feature already noticed in Bavaria but looking rather different in Scandinavia (where it is quite common) owing to the larger wheel. The wheel uprights are at right-angles to the stock, and stays leading from the stock screw into them for wheel adjustment.

A variation featured particularly on Swedish spinning wheels – though also found in Finland and other Scandinavian countries – is the use of double wheel uprights and no stays. On each side of the wheel the two uprights are joined by a semi-circular piece of wood to form an arch, in the top of which a groove is cut to take the wheel axle. This is held in place by a cap-shaped knob. The double upright and arch remain a feature of Swedish and Finnish spinning wheels still being made. When there came a revival of hand spinning in England at the beginning of this century, it was to Norway and Sweden that spinners and

44 Spinning wheel from Norway (probably Telemark) with truncheon-shaped distaff. *Photograph: Norsk Folkemuseum, Oslo*

weavers turned for tuition and equipment, these being countries where there had been no break in the tradition of making textiles for the home by hand. As a result Scandinavian spinning wheels were imported and copied and are still amongst the most popular in use today.

It is well known that it was to Holland that in 1632 Thomas Wentworth, later Earl of Strafford, sent for spinning wheels to encourage the linen manufacture in Ireland in the interests of England's jealously guarded wool supremacy. The industry was further revived in the reign of Charles II by the Duke of Ormonde when he was Lord Lieutenant of Ireland and he is considered to have had more success than Wentworth in getting the people to use the spinning wheels. In Ireland the flyer wheel is called 'Dutch' to this day; alternatively it is known as Old Irish or Low Irish, the latter denoting that it was used in a sitting position. There is no reason to suppose that the Irish spinning wheels have altered in appearance since their introduction and those that still exist show little or no change since they were portrayed in William Hincks' engraving of 1783 (37). Wheel diameters average 19–20in. (48–51cm.) and rim depths 4in. (10cm.); driving bands were often made of plyed wool, sometimes of as many as nine single threads and here again gut was sometimes used. These spinning wheels used to be for flax spinning but more recently they have been used for wool.

'Hugh Macneil Oge lamented that the invading Scots had carried off numbers of spinning wheels from the people of Ulster during one of their depredatory incursions against the North Irish' – so writes Hugh McCall (1855), and certainly examples from the Scottish Western Isles, Harris, Benbecula and St Kilda are almost identical with those in Ireland. Their style is simple but heavy-looking, and an example from St Kilda preserved in the Art Gallery and Museum, Glasgow, is stained a red colour, probably with the same stain that was used to mark sheep after shearing. The wheel diameters are 19–21in. (48–53cm.). Smaller horizontal wheels were used also in the Western Isles.

The first flyer spinning wheel to arrive in the Shetland Isles was also horizontal and is known as the Norrawa, indicating that it came from Norway, with which the Shetlands had long connections. For wood, some of the islands used to rely on wrecks, and it will often be found that while the driving wheel is beautifully though simply made out of a good piece of wood, probably oak, the rest is made of driftwood or other bits that could be found.

Of course spinning wheels were extensively made in many parts of the Scottish mainland, the wheels generally having neat, clean lines with simple turnery, even those made for the more well-to-do. An Act of Parliament of 1751 (24 Geo. II para. XV) '. . . for the better regulations of the linen and hempen manufacturers in that part of Great Britain called Scotland; and for further regulating and encouraging the said manufacture. . . .' decreed 'that every maker of heckles, wheels, reels, weaving-looms and weaving reeds, shall mark or cause to be marked, with an iron brand, or some other proper instrument, in legible and durable characters . . . his christian name, surname and place of residence. . . .', and penalties were laid down for not obeying. Many of the Scottish names were written as pronounced, so spinning wheels can be found marked Piter (Peter), Broon (Brown), and so on. In most countries it is possible from time to time to find spinning wheels which have the maker's name or initials stamped or carved on the stock (this seems to have been particularly prevalent in the nineteenth century in North America) though in Europe by no means as a regular practice. On the better-class ones it is sometimes engraved on a metal, silver or ivory plaque attached to the stock or to an upright, though more discreet places were sometimes found. But because of the Parliamentary Act far more Scottish ones are marked, and their wheelwrights continued this tradition even when they emigrated.

It is very often easier to recognise the social class of home in which a spinning wheel was used than the locality of its origin, and this is true of England where so many of the hand textile tools have now vanished. The Industrial Revolution cannot be entirely to blame since it is in the North of England that more spinning wheels have survived, at least in public collections. This is not surprising when it is remembered that it was not only wool and cotton that were spun here, but that the North was also one of the English flax-growing areas.

Spinning wheels for the more well-to-do were often inlaid or carved, and sometimes the stock was fiddle-shaped, giving added distinction, not that this latter was unique to England, being a shape also met with in Holland and Scandinavia. However, it would appear that in England the hoop rim persisted among flyer spinning wheels made for the ordinary person, and this may to some extent account for the disappearance of so many, since such rims, though cheap to make, easily broke, unlike those made with felloes. When, in the last quarter of the eighteenth century, Romney painted Lady Hamilton sitting at a spinning wheel, it was no elegant wheel that she used. The idea, we know, was Romney's inspired by seeing a cobbler's wife spinning in the market. To fit in with the rustic setting he obtained for the pose a wheel with a hoop rim. Nothing better than spinning could show off those beautiful hands; the distaff with the flax wound round it is on the left and the lady is using only her left hand to spin with. So convincing is the hand position that, with the wheel depicted actually in motion, one could well believe she was herself practised in the art.

It is a spinning wheel with a hoop rim which is the only one in England, as far as one can tell, that has a local name: the Huddersfield broad wheel, with a rim $3\frac{5}{8}$in. (9·2cm.) wide and a diameter of $21\frac{1}{4}$in. (54cm.); an example has survived at Cliffe Castle, Keighley, Yorkshire. Also still in existence are the spinning wheels of two famous men in the history of the early spinning machines, Richard Arkwright and Samuel Crompton. Both men were originally from Lancashire, and although the two wheels are quite different they probably both date from the middle of the eighteenth century. Arkwright's is now housed in the Science Museum, London (45). The wheel is made with felloes and an unusual feature is that the mother-of-all is not the usual disc, but a turned post raising the whorls very close to the wheel. The knob turnery could suggest the Cromwellian period, nearly 100 years before, but it is not known whether this spinning wheel was a family heirloom; its type of turning is found on spinning wheels even among those known to have been made in the nineteenth century.

Samuel Crompton's spinning wheel (46) can be seen in the Hall-i'th'-Wood Museum just outside Bolton in Lancashire, which was Crompton's home from 1758 to 1782 and where he invented his spinning machine commonly called the 'mule'. We know that this spinning wheel belonged to his mother. The wheel is of the hoop rim type with elaborately turned spokes tapered towards the hub to be more weighty at the rim and give momentum to the wheel. As well as a treadle, there is a handle which was used for putting the wheel in position before spinning commenced (*Encyclopédie*). This was quite common on eighteenth-century wheels and may mark a tradition left over from days before the treadle was added. The mother-of-all on Crompton's wheel is a cross-bar without a disc, since the tensioner, instead of entering into the stock and screwing into the part of the mother-of-all which projects into the stock, goes through two vertical chocks attached to the stock, one on either side of the mother-of-all, which is screw-threaded to receive it. A pin at the end of the tensioner prevents it from being unwound too far. The distaff is a straight post as sometimes found again in Lancashire. The stock, probably oak, has a pattern of small circles indented round the side. Instead of metal hooks the flyer has small round wooden pegs. To the front of the mother-of-all is a tracker screw with a pear-shaped handle,

45 Richard Arkwright's spinning wheel with
a lantern distaff. Wheel diameter 19in.
(48·5cm.).
*Photograph: Crown Copyright, Science
Museum, London*

46 Samuel Crompton's spinning wheel. Mid eighteenth-century, wheel diameter 15 in. (38cm.).
Photograph: Museum and Art Gallery, Bolton

matching the tensioner, which can be turned to adjust the mother-of-all to align the whorls with the wheel.

In the Welsh Folk Museum, St Fagans, near Cardiff, there are several examples of nineteenth-century spinning wheels similar to the Lancashire one of Crompton's but less sensitively made. These wheels are sometimes referred to as Welsh wheels though it seems likely that they were originally from England and were adopted and survived in Wales; some of these spinning wheels have flyers reminiscent of German flyers, being rather bulbous in the centre and with arms of separate pieces of wood. Although the wheels have hoop rims, some are lightly gouged round the circumference to guide the doubled driving band. Occasionally they have a T-shaped tensioner handle, a feature which can also be seen on Scottish horizontal wheels and some in Brittany, though in both these two regions the driving wheels are made with felloes.

HORIZONTAL SPINNING WHEELS WITH FRAME

In France, except for the Picardy wheel and those just mentioned from Brittany, horizontal spinning wheels are mainly built into a frame instead of having the solid foundation of a stock. The *Encyclopédie* gives, circa 1765, a detailed descrip-

47 Illustration of a French spinning wheel, distaffs, spindles, niddy-noddy, reel and hank holder. Diderot's *Encyclopédie* 1765

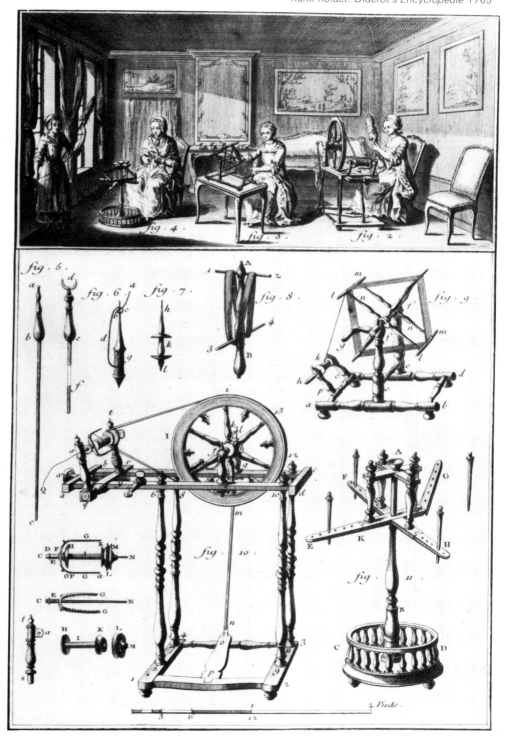

tion of this spinning wheel and how it should be used (47). The construction is based on a square floor frame with four turned upright posts which support a rectangular frame holding both the driving wheel and the flyer mechanism and taking the place of the stock. Two short uprights support the wheel, which is quite small, with six spokes and with points between them on the inside of the rim to decorate it and give it more weight. The footman joins the wheel axle at the back on the outside of the horizontal bar of the upper frame; inside the floor frame it joins the dainty little treadle which is hinged to the front cross-bar of the frame. To the left of the spinner is the mother-of-all, in French called *la coulisse*, 'the slide'. *Les marionettes* 'puppets' (not 'dolls' as on the Picardy wheel) is as delightful a name as the English equivalent, maidens, while the flyer is called *l'épinglier*, literally 'the pin-bearer'.*

The spindle whorl is slightly cone-shaped to reduce the friction against the leather bearing and there is a doubled band drive. The tensioner, with a little knob for a handle, screws through the bar and through the *coulisse* so that when the tensioner screw is turned the whole flyer mechanism slides along the sideways extension of the frame. One of the features of French spinning wheels is the length of the frame on the flyer side which allows the *coulisse* to be positioned, if desired, quite some distance from the wheel. Along the front horizontal bar is a small protruding arm for holding the distaff; it is held in place with a small screw-peg and therefore can be moved along (or even removed). Sometimes these spinning wheels have a levelling screw beside the base of the front right-hand leg, since with four legs they are not so stable on uneven floors as three-legged spinning wheels.

The actual technique of spinning given in the *Encyclopédie* describes drawing the fibres with the right hand. Distaffs on French wheels are often on the right of the orifice (if not in an arm, as above, in a plain hole in the front horizontal bar), but paintings and photographs of French spinners in fact usually show the distaff removed from its holder and held across the lap; the fibres can then be drawn-out by either hand.

An attractive and often-seen variation of the French spinning wheel has a

*Herein may lie the explanation of how the Sleeping Beauty came to prick her finger on the spindle. As in other fairy tales coming from the Continent the fibre is flax, and although there is the suggestion that she was pricked by the sharp point of the spindle – the word *la broche* meaning both spindle and needle – on the spindle wheel, yet this was unlikely to have been used for flax spinning in the home. An early version of the story traced by Ion and Peter Opie in *The Classic Fairy Tales* (1974) tells how when a Neapolitan princess took hold of the distaff a splinter of flax (i.e. a piece of remaining boon) ran under her finger nail. This sent her into a deep sleep from which she was awakened by her own child sucking out the splinter. When, at the very end of the seventeenth century, Charles Perrault, a member of the Académie Française, wrote his *Histoires ou Contes du Temps Passé* (1697) to amuse his children, he changed splinter to spindle – or rather this was how it was translated into English. Perrault may have thought that the Princess pricking her finger on the pins of the flyer (or it could have been a single pin or thorn moved from hole to hole) made a more convincing picture for his children who, no doubt, by then were familiar with the flyer spinning wheel for spinning flax.

triangular floor frame, not a square one (48). This may be a development from the early nineteenth century for it appears in the drawings of spinners at work by the nineteenth-century artist Jean François Millet. His drawings further show that although these wheels have graceful lines and simple but attractive turnery, they were not just drawing-room models but were used equally by the working people.

It could reasonably be said that these spinning wheels with either a square or a triangular base have been used in most parts of France, and also reached the French-speaking part of Belgium. Although the graceful style is the same throughout, these spinning wheels are made in different sizes, their proportions always nicely balanced. A certain number of these French spinning wheels are present in English collections, though whether imported or copied is not known.

Rettich illustrates a horizontal spinning wheel built with a square floor frame, four slightly splayed vertical legs and a rectangular frame supporting the wheel (diameter 18in., 46cm.) and flyer mechanism. The turnery is ornate and the distaff, a single turned post, is mounted to the left of the flyer by means of an arm attached underneath the mother-of-all. In its construction it is similar to the French spinning wheel but the elaborate working of the wood and a certain compactness, the flyer being closer to the wheel, completely changes its style. There are also other differences from French spinning wheels, the most important being that it is used with a flyer drag, and another being the construction of the treadle. The treadle bar runs from front to back of the floor frame on the left-hand side, with the treadle itself at right-angles to it, parallel with the plane of the wheel, and the footman joins the crank at the front of the wheel outside the horizontal bar. Though Rettich ascribes this spinning wheel to Austria (and states that it is alleged to be 200 years old) it is of a type which Horner calls strictly Tyrolese. This is confusing since in Germany and Italy it is said, as Rettich himself says, that it is the vertical spinning wheel which is named after the Tyrol – though Horner calls this 'German'. The use of countries' names to describe types of spinning wheels is misleading when the same or similar features can be found elsewhere. Certainly it is in both the Tyrol and southern Germany that the flyer drag is so prevalent on both horizontal and vertical spinning wheels, the former being of the frame type of which we are now speaking. Even if it were said that it was this type of drive that acquired the name of Tyrolese, and not the type of construction, it could easily be disproved.

The skill of the wood turners and carvers in the Tyrol produced some very fine ornate spinning wheels, the woods being mainly birch, pine and walnut (49). In the example shown the treadle lies in the same direction as on the French wheels. Another variation in the Tyrol is where the four slightly splayed legs from the floor frame carry a flat platform from which four short posts rise to support the horizontal frame; some examples have a gallery round the platform. However horizontal spinning wheels with frame are not confined in Austria only to the Tyrol, being found also in other parts of the country.

In Germany similar wheels can be found in Swabia and Bavaria; some have very small wheels, approximately 12in. (30·5cm.), and a stylistic variation is a diagonal cross of turned posts between the two back legs, floor frame and cross-bar, for strength and stability. Simple types of these spinning wheels without extravagant turnery but with larger wheels, approximately 17in. (43cm.) in diameter, and with the wheel axle resting in nothing more than a block of wood, were used up until quite recently for spinning wool (**50**).

In Switzerland this same type of spinning wheel was widely used in the canton of Graubünden for spinning flax. Another, coming from south-west of Basel, again has a drag on the flyer, though being so close to the French border it might have been expected to have had two whorls and a doubled band drive.

48 French spinning wheel with triangular-shaped base. Wheel diameter 15¾in. (40cm.).
Photograph: Pitt Rivers Museum, University of Oxford

50 Horizontal spinning wheel with frame and lozenge-shaped distaff from the Tyrol. Wheel diameter 14½in. (37cm.).

Photograph: Tiroler Volkskunst-museum, Innsbruck

51 Vertical spinning wheel from the Tyrol with an X-shaped base, and flyer drag. Wheel diameter 12½in. (31·5cm.).
Photograph: Tiroler Volkskunst-museum, Innsbruck

49 Horizontal spinning wheel with frame in use for spinning wool in the Bavarian Alps.
Photograph: Deutsches Museum, Munich

In Italy the drag on the flyer was made by wrapping a piece of rag around the neck of the flyer with the string passing over it as already described (page 79). It would seem that the only type of horizontal spinning wheel found in recent times in Italy is this frame type, in the east scattered in the Venetian plains and in the mountains on the borders into Austria; in the west, in the frontier regions of Piedmont, while at the turn of the century Horner found it in both Lombardy and the Tuscan Apennines.*

Moving east to Czechoslovakia and Hungary one can find a simplified version of the horizontal frame wheel: the upper frame does not enclose the wheel but ends just beyond the axle which is held in slots in the upright; thus the wheel projects well beyond the frame. The wheel is approximately 22in. (56cm.) in diameter and the tensioner is off-centre towards the front, in order not to interfere with the movement of the driving wheel. An example from Hungary in the Horner collection has a doubled band drive. Horner himself describes it as a most unstable wheel.

In Scandinavia, there are some spinning wheels which could be described as a mixture of the two horizontal types: a square floor frame and four slightly splayed legs support a horizontal stock, slotted to take a large wheel (approximately 24in., 61cm. diameter) which again extends beyond the frame. In Denmark there are spinning wheels which have three legs and a low horizontal stock, with wheel uprights that reach above the wheel where they are connected by a cross-bar. (A similar feature was thought of by a spinning-wheel maker in the French Hautes-Alpes.)

In Britain, spinning wheels with a frame instead of a stock and with the flyer drag are rarely found; of those that are, most are known to have been brought from Austria or Germany. In their traditional or true form, the type seems to belong to the regions in and around the Alps, although certain features of them appear in countries to the east and north.

VERTICAL SPINNING WHEELS WITH BASE

The vertical spinning wheel with a stool base is as widely spread through Europe as the horizontal wheel with stock. The important difference between the two is that on the vertical spinning wheel the flyer, spindle, bobbin and its supports, maidens and mother-of-all are placed higher than the wheel. According

---

*The *Dicziunari Rumantsch Grischun* and *Bauernwerk* by Scheuermeier give excellent line drawings of this frame type of spinning wheel, in both cases with the wheel on the left and the flyer mechanism on the right – an interesting difference, on the face of it a speciality of northern Italy and Graubünden, but none of the photographs or museum examples have this feature, so it must be taken for an artist's error. Apart from one or two spinning wheels obviously wrongly assembled, I have so far come across only one horizontal type built with the driving wheel on the left and this mounted on a stock not in a frame. This spinning wheel is thought to have come from Austria or Germany, and from its elaborately made wheel and its flyer drag it probably does so. The driving band is of leather, another feature more prevalent in Austria, Germany and Switzerland than in most other countries.

to where they were made, any of the three types of drive were used. Inevitably the driving wheels were kept small, seldom exceeding 16in. (40·5cm.) in diameter and usually less, since if they were large it would raise the flyer too high for the comfort of the spinner. Vertical spinning wheels appear to have been produced in the main for spinning flax, for which they are well suited. Smaller wheels require faster treadling but if this is kept even it need not be a drawback for the regular rhythm of flax spinning. When well made and with distaffs attached they can be very handsome pieces of equipment, although tall or heavy distaffs can sometimes incline them to overbalance in use. Their great advantage is their compactness: it is easy to imagine how they may have been developed from the horizontal type to take up less space and be less vulnerable to breakage.*

The stool bases vary in shape: rectangular or six-sided; fancifully curved, or heart-shaped, probably indications of an engagement present; or sometimes completely round though often with a 'bite' out of the circle to allow the spinner's foot to reach the treadle comfortably. For the same reason the legs, two in the front and one at the back (occasionally two at the back) are splayed, bringing the treadle well forward. In countries as far apart as Scotland, Sweden and Belgium the base is not solid but composed of two parallel battens joined at either end by short transverse pieces, making a space through which the lower circumference of the wheel comes below the level of the base. This prevents the orifice becoming positioned too high, even if a fairly large wheel is used. Another kind of base, made in the Tyrol, is of two bars crossed in the form of an X with a leg at each corner (51).

In Yugoslavia there are T-shaped bases raised on three legs, one at each end of the T. These spinning wheels are constructed in two ways. In one arrangement the arms of the T face the spinner, the treadle bar being placed between the two legs below these. In the other arrangement the treadle bar runs obliquely from beneath the right-hand arm of the T to beneath the stem of the T. The mounting of the wheel and flyer mechanism is now reversed, and the whole spinning wheel swung round, bringing the orifice to point obliquely to the spinner's left. This is an appropriate and comfortable position for spinning flax with one hand, or for plying.

Normally on vertical spinning wheels the treadle bar fits between the two front legs and a single piece of wood, the treadle, about 4–6in. (10–15cm.) in width, is attached at right angles to it and narrows at the end where it is tied to the footman. Some are very obviously ornamentally-shaped and some have heart-shaped cut-outs of clear intention. Occasionally, as with examples in the Brunswick Landesmuseum, the treadles are in three pieces like those on horizontal wheels. Decoration in ivory or bone, and elaborate turning or carving are as

---

*In Belgium and Holland the name given to the vertical spinning wheel is *schippertje*, or barge wheel. This name derives from the time they were used on the barges: when there was no breeze the man would tow the boat, and the steady motion then made it possible for the woman to spin. The neat little wheel could be fitted into a barge's cramped quarters.

varied as on horizontal types, but one form of decoration used in the Tyrol and northern Italy can be simply by little roundels of paper stuck on the sides of the wheel or stock to form patterns.

There is an economy of design in vertical wheels in that the two main uprights from the base support both the wheel and the flyer mechanism. Forms of tensioner vary: one type requires the continuation of the uprights above the wheel to be shaped evenly round and smooth and joined at the top with a cross-bar. The mother-of-all is drilled with a hole at either end and can slide up and down the smooth pillars when moved by the screw tensioner which is centred on the cross-bar with its handle above. (Such a tensioner is shown in plate 27 on page 79 although in this example there is a second supporting front upright.) This form of tensioning occurs mainly in areas which also have the horizontal wheel with frame, the principle being exactly the same in both, though in the vertical wheel the tensioner is above the wheel instead of alongside it. It is found thus in the Tyrol, parts of Switzerland, northern Italy in Trentino and Piedmont, where it is known to have been made locally; also in central and south-west France, though here, surprisingly, usually with a drag on the flyer, unlike the French horizontal frame types. An example in the Musée des Arts et Métiers in Paris is described as being of the Louis XV period and has, instead of string as a friction band, a piece of wood fitting round the hollow shaft and is tightened with a little peg.

In Germany the same type of tensioner is met in the Black Forest and Wurtemburg as well as Bavaria, but it is in Germany that a rather different type of tensioner can be found, one which is also met in Hungary, Czechoslovakia and Romania. This tensioner is built into an extension of the back upright, the handle passing into the top and the screw going through the mother-of-all in a cut-out slot in the upright. With this kind of tensioner the maidens are positioned vertically, the back one close up against the tensioner, so that the spindle always remains horizontal when moved. A cross-bar just above the wheel joining the two main uprights keeps the construction rigid, but even so it leaves the front of the mother-of-all unsupported. A wheelwright in Bavaria overcame this by extending the cross-bar and adding a wooden screw, with its handle underneath, to go through the bar and support the mother-of-all on its point.

Yet another form of tensioner is found in Central Europe and Scandinavia, and also in England, Scotland and the Low Countries. On such vertical wheels there is no back maiden. This changes the role of the mother-of-all since it is no longer moved by means of a tensioner; instead it forms the cross-bar which ensures rigidity between the two uprights, and extends forward to support the vertical front maiden. The other end of the spindle beyond its whorl fits into a small chock (or sometimes just lies in a groove on top of it) which is moved up and down by the tensioner in the slot in the back upright. The slot can be made either from front to back or else from side to side, in which case the block will protrude slightly to the right to support the spindle outside the upright in alignment with the bearing of the front maiden. In either case this means that all the tensioner does is to lift or drop the end of the spindle, which is therefore no longer kept horizontal or so steady. One way of overcoming this weakness, seen in England and Scotland, is to have the front bearing cut with two or three holes in the leather so that the orifice can be moved to keep it more or less level with the end of the spindle.

In Spain, vertical spinning wheels have been used both by the well-to-do village folk and middle-class people in the north, north-west and the Basque provinces, and are known to have been made locally. One preserved in the Museo del Pueblo Español, Madrid has the last type of tensioner described above and a doubled band drive.

52 A vertical spinning wheel of a type used by the bourgeoisie in the nineteenth century, in Copenhagen.
Photograph: Nationalmuseet, Dansk Folke-museum, Brede

A fashion in spinning wheels at the beginning of the nineteenth century was for cabriole legs. They are seen on French and German vertical types. In Denmark the bourgeoisie of Copenhagen adopted a pretty little vertical spinning wheel at a time when it became fashionable to spin flax, and some of the existing examples also have cabriole legs. The driving wheels of the Danish ones are made of brass with turned wooden spokes, with diameters of 14–15in. (35·5–38cm.) and a curved brass distaff arm attached to the back upright adds an overall elegance. They may have been German influenced since some of them have a front maiden which is cup-shaped and grooved to take a drag across the flyer shaft. However, a photograph of one in use shows a doubled band drive and signs of unsteadiness caused by inadequate support for the front maiden; to help overcome this a string has been tied between the back upright and the maiden (52). In this particular instance the flax would appear to be looped round with the two ends knotted at the top and the whole bundle kept hitched up by the wide band. The girl spins from a film of fibres drawn from the bottom of the loop. These wheels also reached Sweden and Norway, though it is not known whether they were made there or merely imported.

In north-east Scotland a rather rare type is built on to a three-legged stool with cabriole legs, and on this account probably dates from the nineteenth century; such legs in fact were needed to give stability to this particular structure (53). The 'boomerang' supports curve round to the left, levelling out at the end and a cross-bar between them holds the tensioner. The support for the flyer mechanism is all in one piece, with leather bearings on top of the upward curves (thus forming the maidens) to hold the spindle. The flyer has holes instead of hooks and one small wooden peg. To hold the wheel axle a short support branches off each main support, curving to the right, where it is slotted. A similar design can be found in Skåne in the south of Sweden but more crudely and simply made and used as a bobbin winder. Here the base is a rectangle standing on four legs and the treadle is similar to those on a horizontal type. The wheel, diameter 24in. (61cm.), has turned spokes. It runs in two slots in the main supports, placed at the point where these curve sharply to the left (they are set into the base tilting slightly to the left). A few inches before the end, where the supports are almost horizontal, are slots to hold the bobbin.*

In Scotland, vertical wheels were made not only for flax but also for wool spinning. It is difficult to make a clear definition between the two, but flax wheels are usually provided with a distaff, and as the thread was required fine, the orifices could be small, the hooks on the flyer close and fine, and since flax

*Both the wheels described have the flyer beside the wheel, not above it, again a mixture of styles. An example from Denmark has the features of a vertical spinning wheel with base, but with the uprights and tensioner bearing the flyer mechanism on the left, and therefore a second pair of uprights to support the wheel. On the other hand a horizontal spinning wheel with frame from the Tyrol has the flyer mechanism positioned higher than the wheel, very close to it on the left-hand side. Though they stem from different types, the relative positions of flyer and wheel amount to much the same.

53 A spinning wheel from north-east Scotland with boomerang supports. Wheel diameter 15½in. (39·5cm.).
Photograph: National Museum of Antiquities of Scotland, Edinburgh

for general household use is better well-twisted the whorls can be closer in circumference. It is known that the Scots had several sorts of wheel, for hemp, worsted, lint, wool, fine woollen spinning etc. In Shetland, where very little flax was grown, a small vertical spinning wheel known as the 'spinnie' largely replaced the 'Norrawa' wheel about 1800. In Scotland gentlewomen considered it their duty to spin the linen yarn for their home requirements – a tradition that lasted well into the nineteenth century – and some very fine spinning wheels were made. Many of them are vertical with elegant distaffs, and, being Scottish, they were never over-fussy (see plate 33). Some had iron spokes with wooden rims.

The flyer drag was not a form of drive much used in Britain but from time to time spinning wheels are found to have it. Usually they are of the vertical type and one would suspect them of being German, either imported or copied. In Wales, flax was very little grown except on certain estates, mainly in the Conway Valley in the north, where it was used by the families that grew it. Two spinning wheels in the Welsh Folk Museum have the characteristics of German and Austrian vertical spinning wheels and may have been imported by owners of such estates, either because they had become fashionable, or because they were considered more suitable for flax spinning, particularly if the local wheel-wrights and wood-turners were accustomed only to making great wheels for the wool spinners. One of these spinning wheels was obviously made for hard work since it has brass shodding at the ends of the footman as well as brass inset into the wheel rim, although it is fashioned out of mahogany.

Horner, Born and Patterson all illustrate as an example of an English spinning wheel a small vertical wheel with curved arches instead of maidens. The front bearing is a leather wedge which slots into the top of the arch and is easily removable for reaching the spindle and bobbin. The arches certainly give it an individuality among vertical types, and there is reason to believe that this arched type was made in north Lancashire. It must have been produced in some quantity, since a fair number are still in existence both in public collections and in private hands (and although seen in many parts of the country, they always relate back to the north-west). All are to the same basic design although details of turnery and treadle shape vary and wheel diameters run from $14\frac{1}{2}$ to $17\frac{1}{2}$in. (37–44·5cm.). However it should not be considered typical of English wheels, but a style adopted by an individual maker (or perhaps by several of them locally). The wheels are made with felloes, but an example in Snowshill Manor with arches has a hoop rim (diameter $17\frac{1}{2}$in., 44·5cm.) and was undoubtedly made by a different maker. Most English vertical spinning wheels, however, have the usual upright maidens, doubled band drive and wheels made with felloes. Another style which comes from the North of England has the wheel uprights made from a single fluted pillar split down the middle to form two pieces, their flat parts facing the wheel. This is quite a common technique for making matching uprights on spinning wheels, not only in England; such fluted pillars are also found on Yorkshire grandfather clocks of the late eighteenth and early nineteenth centuries, and the style may have been echoed by a spinning wheel maker.

VERTICAL SPINNING WHEEL WITH FRAME

Several different types of spinning wheel are to be found in Bavaria, amongst which is the vertical wheel with frame. It is one of this type, marked with the date 1604, that was mentioned earlier as being possibly one of the oldest preserved spinning wheels with treadle. Since this type in no way resembles the early horizontal spinning wheel, its development may have been influenced by some quite different piece of equipment which had a wheel drive turned by means of a treadle, as for grinding, drilling, polishing or turning, but there is absolutely no positive evidence that this was so.

This vertical spinning wheel has a rectangular floor frame, the two longer bars being more substantial than the two cross-bars since they carry the weight of the wheel. Two flat uprights are attached to the floor frame, opposite each other, and are notched to hold the wheel axle just high enough for the rim to clear the floor. The diameter of the wheel is greater than it is on other types of vertical spinning wheels. To keep the wheel hard up against the right-hand upright, the axle is a rounded wooden bar about 6–8in. (15–20cm.) long which terminates at the left-hand upright, where a U-shaped crank is attached to it from the outside. This in

Joséphine, Lydia, Hélène... et leurs moutons

Fidèles à une tradition transmise de mère en fille, les paysannes de Grächen, les vraies, celles qui fument encore la pipe, tondent, cardent, filent et tricotent la laine de leurs moutons.

54 Vertical frame spinning wheel used for flax spinning in Bavaria from a free-standing distaff.
Photograph: Deutsches Museum, Munich

turn is hooked on to the short footman, the treadle being also outside the frame. A bar bridges the tops of the uprights and underneath it the mother-of-all fits into slots cut in the inner side of the uprights. On the German ones there are generally two tensioners which must be turned together so that the mother-of-all keeps horizontal as it is moved up and down in the slots. The maidens are horizontal, so they are not directly above the wheel as they may be on vertical types with base. The spindle fits into a hole near the end of the right-hand maiden (equivalent to the back one) and rests in a notch in the front maiden. The drag is on the flyer and there is one groove in the wheel rim. The flyer is of metal and has notches instead of hooks on the lower left-hand side and upper right-hand side. Some of these spinning wheels are ornately carved, including captive rings on the cross-bars, and are highlighted with paint. For the spinner to reach the treadle the spinning wheel is placed beside her on the right, which brings the plane of the wheel virtually at right-angles to her knees with the flyer orifice facing across to her left. It is usually treadled with the right foot so that the left can steady a free-standing distaff when flax spinning (54). But on some of these spinning wheels the cross-bar beside the treadle is well worn, which shows that the right foot must have rested there while the left treadled. The whole design was surely intended for using with the left hand to draw the fibres (and if one hand only were used, also to control the twist) and of course – once again – may have also been used for plying or twisting. It is also likely that it was used more often to give an S twist because of the position of the notches on metal flyers although on some examples which have wooden flyers the hooks are on the upper sides of both flyer arms.

The vertical frame wheel is often referred to as the Swiss wheel, or even more specifically as the Bern wheel, but unless it is truly an indication of where it was in fact first made, this use of place or country names is again misleading when the type of wheel is known to have been used in other areas. Horner found this type of wheel in Switzerland from Bern to Ticino, though it seems to be considered rather rare in Graubünden, where there are some differences in construction from those in Germany and for that matter from other parts of Switzerland (55). Instead of the elongated wheel axle, the wheel is mounted in its own frame and the treadle is inside the main frame. A cross-bar above the treadle gives rigidity, but with only a single tensioner (which is the norm in Switzerland) the mother-of-all does not necessarily remain absolutely horizontal within its slots. The flyer is of wood and has holes, and the peg for the drag fits into the end of the maiden. The *Dicziunari Rumantsch Grischun* records one dated 1699, from the eastern part of the canton. It would seem that the date quite often appears on these types of spinning wheel, usually inscribed on the mother-of-all or on a bar above it.

Six examples in the Museum für Volkskunde in Basel, one dated 1821 and another 1838, all come from the canton of Bern and resemble each other but are different from the one just described from Graubünden. The driving wheels are

55 Vertical frame spinning wheel from Tschlin, Lower Engadine, Switzerland. Wheel diameter approximately 21in. (53cm.). *Photograph: Dicziunari Rumantsch Grischun*

large, with a diameter of 25in. (63·5cm.); the treadle is outside the main frame, attached to an extension of the front cross-bar. All have two grooves in the wheel rim and therefore a doubled band drive. The spindle rests in recesses in the horizontal maidens, and a thin batten of wood, of the same width as the maiden, is attached at one end of it so that it can swivel over the top of the orifice to keep it in place, or be turned to release the spindle for reaching the bobbin. Attached to the batten is a small wooden cup for keeping lubricating grease for the bearings and axle, or possibly water for flax spinning, though in more recent times this type of wheel has been found used for wool. In the village of Brientz in 1973 I saw it in use also for spinning silk waste.*

The vertical spinning wheel with frame is also found in France and, not surprisingly, towards Switzerland in the Savoy Alps. Two examples (with pro-

*In the eighteenth century when sumptuous clothes at reasonable prices were in demand by the expanding middle classes, the spinning of floss or silk waste was a cottage industry in Switzerland. The raw material came from France and Italy (*Ciba Review* no 111).

venance not stated) in the Musée National des Arts et Traditions Populaires, Paris, both have flyer drags.

Rettich illustrates a spinning wheel which he calls 'Otzthaler Spinnrad, the newest type of wheel from 1893'; one may find it hard to believe that it took this long to reach the Tyrol. This Austrian version has the treadle outside the frame and a flyer drag. It also has a distaff and an articulated distaff arm screwed into the top cross-bar beside the tensioning nut. In this instance the tensioning screw is attached to the mother-of-all and the screw-thread goes upwards through the cross-bar; raising or lowering is by the tensioning nut. In 1904 Horner found this type of wheel being made by one Josef Auinger at Prambachkirchen, Upper Austria.

The vertical spinning wheel with frame certainly moved further east, since in Poland, in the Ethnographic Museum in Krakow, there are two examples from a farmhouse in the region, both with flyer drag and the treadle inside the frame.

It is often unexplained how a type of spinning wheel reached a different area. One possibility is with a bride moving to her husband's home, and another is through spinning-wheel makers themselves moving about, for instance if conscripted into military service. But sometimes, one feels, similarities have been arrived at quite independently.*

In Italy, according to Scheuermeier, the vertical frame spinning wheel has been used near Aosta, where it is considered to have been influenced by the German-speaking part of Valais, and also in Domodossala, where it was made for the first time around 1860 by a German. The Italian versions sometimes have a distaff attached to the top of the spinning-wheel frame for flax spinning, although distaffs are here often tucked under the arm. When more recently the wheels have been used for spinning wool, this has been spun from the lap. Usually there is a doubled band drive. The treadle can be either inside or outside the frame.

There is another type of vertical frame wheel which is even simpler than those so far mentioned, but this does not necessarily mean that it is older. Its feature is a triangular-shaped frame instead of a rectangular one. The three-sided floor frame bears a stout upright at one of the corners, which supports the wheel (diameter approximately 15in., 38cm.) so that this just clears the floor. Above the wheel at right angles to the upright is a cross-bar which, as in other vertical spinning wheels, supports the front maiden and flyer mechanism, the spindle resting in a tensioning device incorporated into the upright. A second upright, reaching up

---

*In the north of Sweden in the area of Norrbotten, a *tvinnrock* or plying wheel dated 1745 was found and is now in the Norrbotten Museum; it is not unlike the type of spinning wheel that we are discussing, the wheel just clearing the floor and the treadle being on the outside of a slightly curved upright. This must surely be an example of an independent invention by the Swedish maker, and not derived from further south.

to the cross-bar, forms the other upright for the wheel axle, with the crank and footman to the front of it; the treadle therefore lies underneath the cross-bar. The spinner sits facing both the orifice and the treadle, and the batten of the floor frame facing her is in fact the treadle bar. To give some stability to this construction, two posts splay out from underneath the front maiden to form a triangle with the treadle bar. It has a large-sized flyer and a drag on it and was probably used for spinning tow. Scheuermeier found it in Italy – the only flyer spinning wheel type surviving south of the Po and then only in one area, Emilia. This was as far south as he found any treadle spinning wheels: he adds that in south Italy they are quite unknown. Horner, however, collecting at the start of this century, procured an example of this triangular type from the Tuscan Apennines, whereas in this part of Italy Scheuermeier could find no more than a mention of treadle wheels.

Vertical spinning wheels with frame are not found in Britain save for the few that can be seen in our museums or in private hands, and these are believed to have come from Switzerland, some with nineteenth-century dates.

Chapter Six

Some special types of spinning wheels

CASTLE SPINNING WHEELS

Castle wheels appear to belong, in the main, to Northern Ireland and Scotland. It is as yet unknown whether these wheels were developed independently of each other or, if one influenced the other, which was thought of first. From time to time they appear elsewhere; for instance there is one in the Nordiska Museum, Stockholm, but as it also has two flyers it will be described in the section dealing with these. The special feature of them all is that the driving wheel is above the flyer mechanism. (In North America the term 'castle' as applied to spinning wheels has now come to denote all forms of vertical wheel, but in Britain it is only used when the wheel and flyer are arranged in this particular manner.)

The Scottish version of the castle wheel has a small rectangular base with long legs which bring the overall height to 48in. (122cm.) although the orifice is 23in. (58cm.) off the ground. The mother-of-all is fitted directly on to the base with the tensioner underneath the front maiden, while the wheel is raised by long uprights to a position above and slightly to the right of the flyer mechanism. As the distaff upright is on the left-hand back corner of the base, an unusually long distaff-arm is necessary to bring the distaff itself sufficiently forward to facilitate spinning from it (56).

In Donegal (north-west Ireland) one notices that the wheel uprights on horizontal types of spinning wheel are quite vertical and the stock has only the slightest slope, so that the flyer is noticeably lower than the axle of the wheel. Whether or not this has any bearing on the development of the Irish castle wheel is not possible to tell, but in the main it is from this county and also Co. Antrim, an important flax growing area, that they come.

This unusual wheel is suggestive of a child's high chair rather than a castle, and with its wheel out of reach of small hands would certainly have been suitable for a cottage full of children. Its three-sided construction makes it both easy to store in a corner when not required, and very stable, an essential attribute for spinning good yarn.

Its average overall height is 42–46in. (106·5–117cm.) so it stands slightly lower than the Scottish one but the height of the orifice from the ground is about the same (23in., 58cm.) since this is a comfortable height for the spinner sitting on a chair or stool. Driving wheels are generally made of oak with diameters of 18–20in. (45·5–51cm.). An example in the Ulster Museum has 'Ann MacDonnel

56 Scottish castle wheel (distaff and driving band missing). Wheel diameter 20in. (51cm.). *Photograph: National Museum of Antiquities of Scotland, Edinburgh*

Cushendall Glens of Antrim 1774' carved round the front rim. On the sturdy tripod framework, the wheel axle runs between the top of the long back leg and a short upright on a cross-bar placed between the two front legs. Also on this cross-bar, to the spinner's left, is a large wooden knob, drilled with a hole, which holds the distaff. A second hole, bored on the right of the same cross-bar, seems to have carried a second knob to balance the other, though sometimes it is left empty. The distaff itself is usually the pear-shaped cage type. The mother-of-all and flyer mechanism are directly under the wheel, supported by a cross-bar jutting out from the back leg. The bearings that support the spindle are on top of the maidens and the flyer has an unusually wide span, as much as $6\frac{1}{2}$in. (16·5cm.), although bobbins average only $2\frac{3}{4}$in. (7cm.) in length and therefore cannot have been filled to such a width. The maidens and a small post in the centre of the mother-of-all extend downwards into the cross-bar, and can be slid up and down for tensioning; there is a nut in the centre of the bar for tightening it. In some cases the centre post is screw-threaded and a disc-shaped nut turns on it for adjustment.

57 Irish castle wheel in use for wool spinning
in Co. Antrim c.1910.
*Photograph: Green collection. Ulster Folk
and Transport Museum*

The hooks on the flyer of these spinning wheels are usually on the upper side
of the right-hand arm and the lower side of the left-hand arm for twisting in an
S direction, which strengthens one's belief that they were originally for spinning
flax. However, once this was no longer done by hand the castle wheel, like other
Irish flax wheels, became used for spinning wool and for this latter purpose a
few were still in use in this present century (**57**).

Some of these wheels certainly reached the east coast of North America, and
there they must have been copied, for a description by Ruth Gaines writing in
1941 (*Homespun*) suggests a much more lightly constructed frame than their
stout counterparts in British collections, although Pennington and Taylor (*The
American Spinning Wheel*) state that those made in America have larger wheels
than the Irish equivalents. One would have assumed that the castle wheel had
travelled from Ireland to America with the Irish-Scotch immigrants, but it seems
that it was not necessarily these people who made them there. Perhaps it was
copied through being recognised as a good design of spinning wheel.

TWO-WHEEL SPINNING WHEELS

Flyer spinning wheels with two wheels appear fairly rarely in Europe. The
principle is the same as that already described in chapter II on spindle wheels,

but with flyer wheels the accelerating wheel is usually closer in size to the main driving wheel. Probably their origins were in Eastern Europe.

Rettich illustrates one which, he says, was introduced by refugee Serbs on the Hungarian military frontiers about 1820. (He adds that it must be one of the oldest spinning wheels with flyer, spindle and bobbin, but he gives no further reference.) The spinning wheel in his example is mounted on a horizontal stock; the smaller wheel, placed in the middle, is turned by a handle and drives the larger wheel on the right which is linked to the flyer mechanism on the left by a long driving band. A small distaff on the extreme left end of the stock suggests that it was used for flax or hemp. There are also such wheels in Bulgaria.

Another European double-wheel spinning wheel is illustrated in *Spind & Tvind* (Kunstindustrimuseet and Herning Museum, Copenhagen, 1975). Again the wheels are side by side but here they have the same diameter and are driven by two treadles, one for each foot of the spinner. The wheel on the right drives the hub of the one on the left which in turn drives the single spindle; a small distaff is placed above the left-hand wheel. Liza Warburg tells us in the text that 'it is praised for its speed and stability and, according to notes from the Society for National Handicraft, 1816, was invented by Pastor Ostrup and at least ten have been made in Denmark'.

In an article by Serge Daniloff in the American publication *Antiques* (1929) there is an illustration of a spinning wheel with two wheels, this time one above the other. It was owned by the then Mayor of Lunn, Massachusetts, and according to Daniloff, 'It is very old, having been brought to the Colonies, it is believed, during the early period of the settlement or made here [the U.S.A.] shortly after'. He dates it to circa 1700. This spinning wheel has a rectangular base with four legs and two treadles, 'the countershaft drive making it possible for the spinner to obtain the necessary speed of the flyer without unduly rapid motion of the feet'. The two treadles drive the lower and larger of the two disc wheels, this being linked by a band to the hub of the upper smaller one; a cross-bar between the two uprights supporting the wheels, separates them. Pennington and Taylor illustrate other such wheels made in America with two treadles and two wheels one above the other of varying designs, though their axles are staggered so that the circumferences of the wheels slightly overlap. Spinning wheels with two wheels and two treadles seem to be found more often in the United States.

A design which is unique is the Connecticut chair wheel, first made around the middle of the eighteenth century. Although found by Pennington and Taylor in other parts of the States, the chair wheels have usually been traceable to Connecticut, where they were made by more than one maker. Styles vary; some have square legs and a rather 'cottage' look; some also have a device for adjusting the tension between the two wheels. In the American Museum in Britain (near Bath) an example has all its turned parts smooth and rounded (58). This spinning wheel is built into the seatless frame of a wooden arm chair. The two wheels, one above the other and slightly overlapping, are placed between the arms where the

58 American chair spinning wheel. Mid-eighteenth-century, Connecticut; lower wheel diameter 14in. (35·5cm.); upper wheel diameter 11¼in. (28·5cm.); solid hub of upper wheel diameter 5in. (13cm.).
Photograph: Reproduced by permission of The American Museum in Britain, Bath

seat would be. The mother-of-all is the left-hand arm of the chair and two tall maidens leaning outwards hold the flyer mechanism. By a plain T-shaped tensioner in front of the front upright the mother-of-all can be tilted inwards or outwards, nearer or further from the wheel. The flyer in this case has only one arm; to judge from the neatly shaped stub where the second arm would be, it looks as though the flyer might have been originally made in this economical fashion and not subsequently broken, but pictures of other chair wheels clearly show a flyer with two arms and hooks on each. The flyer mechanism is driven by a doubled band which encircles the upper and smaller of the two wheels. This in turn is driven by its hub from the lower and larger wheel. There are two treadles, each with a footman linked to a zig-zag crank in the axle of the lower wheel. The distaff is attached to the back of the chair and has a pear-shaped cage of willow sticks; the fibres have to be drafted over the top of the flyer to the orifice and therefore both hands would probably be used. The spinner sits directly in front of the chair with the flyer slightly on her left and a foot on each treadle. It is a very stable spinning wheel to use, and fast.

Daniloff, commenting on these spinning wheels, writes: 'The reason for the chair form . . . is not clear. One may venture to suggest that, in early days, old chair frames were, at times, actually employed by the first settlers as skeletons for their wheels. Later, perhaps, the unusual shape caught the eye of a following generation and the original chair frame wheel was duly imitated.' However, chair making and spinning wheel making sometimes went together and of the 300 or so furniture makers in Delaware between 1740 and 1860 listed by Harold B. Hancock (1974) there is mention of 12 spinning wheel makers, five of whom were also chair makers.*

DOUBLE FLYER OR TWO-HANDED SPINNING WHEELS

As it is possible to spin fibres attached to a distaff with one hand only, it is understandable that the idea arose of using both hands independently to produce two threads at once.

The earliest evidence of a spinning wheel built for this purpose is in 1677, when an Englishman, Thomas Firmin, living in the Parish of Aldersgate, London, published in 1681 *Some proposals for the imployment of the Poor and for the prevention of Idleness and the consequences thereof, Begging.* Written in the form of a letter to a friend, he starts by telling him, 'it is now above four years since I erected my work house in Little Britain for the Imployment of the Poor in the Linnen manfacture. . . . At this time I have a person who for five shillings a

*The painting of 1610 by Jan Joris van Vliet of Delft shows a Dutch chair maker at work with a half-made horizontal spinning wheel in the foreground. It may be noted in this connection that the rules of the Guild of Chairmakers of 1713 in St Truiden, Belgium (Jozef Weyns) stated that to 'become a master craftsman the test was not to make a chair, but two pieces, one of which was to be a double spinning wheel with tensioner, 'een dobbel spinnewiel met een schrouff' (with a screw). If this could be interpreted as meaning two wheels, could such test pieces have been the precursors of the American chair wheel?

Shee layeth her Hand to the Spindle and her hands hold the distaffe: Pro: 31. 19.

59 The two-handed spinning wheel illustrated by Thomas Firmin in 1681

week, doth constantly teach between twenty and thirty poor children to spin; some that are little, upon the single wheel, and others that are bigger, upon the double, or two handed wheel (like that which you have at the beginning of these papers, which I esteem the best way for spinning, and full as proper for wooll as flax)'. The sketch that he refers to is of a vertical spinning wheel with two flyers, driven with one doubled band which runs in two grooves on the wheel and over both sets of whorls (**59**). Each flyer mechanism has a separate tensioner and the distaff is positioned between the two orifices. The flax, which appears to be dressed straight down on a tall distaff, is covered with a sheath and tied near the top and twice more near the bottom. It stands to reason that the flax must be very carefully arranged and firmly tied for spinning two threads at once. A sponge is tied on to the distaff arm and so hangs conveniently for wetting the fingers.

Although Firmin mentions that the children in Holland could contribute to their maintenance by the time they were seven or eight years old from the money they earned by spinning, he in no way suggests that they used the double flyer spinning wheel, nor that it came from there. However, the fact that he also considered this wheel appropriate for spinning wool as well as flax may indicate the use of a distaff for spinning a worsted-type thread as the Dutch did, although Firmin's own business was entirely in flax. This was used for making allabas cloths and coarse canvas for pepper bags. 'A child may first be set upon spinning of tow, which cost about two pence a pound, which though it be spun never so badly will serve for some kind of use or other, which in wool would be good for nothing, which yet is many times the price.' Firmin advocates that the poor children, besides learning to read, should learn to spin on both the single and the double wheel. 'But before you enter the child upon the double wheel', he writes, 'or suffer it to spin with both hands, you must teach it to spin well with either hand upon the single wheel, which is turned with the foot as the other is, by means whereof, you may teach the child to draw out the flax with either hand indifferently, and to be as nimble and quick with one hand as with the other; otherwise, when it comes to spin upon the double wheel, which hath a quill for each hand, the threads will not be alike, but one will be more twisted, and the other little, and so will not do well together.' It is not known whether Firmin's ideas were put into practice in other establishments and little more is heard of the double flyer spinning wheel until well into the eighteenth century. It was mostly found where flax spinning was an industry, as it was thought of as a wheel essentially for helping the poor to earn more money and to speed up production although it could not actually double the output.*

Stevenson's report on the linen trade in Ireland in 1755 records only two schools for teaching the two-handed spinning wheel, one in Dundalk and the other in Greenvale.

It was in Scotland that the two flyer spinning wheel seems to have come into its own. In the *Statistical Accounts of Scotland* of the late eighteenth century there are references in both Angus and Aberdeenshire (both in the north-east where much of the Scottish flax was grown) to the general use of the two-handed spinning wheel: '. . . the women here use all two handed wheels as they call them; they are in general capital spinners and bring a deal of money to the Parish', although in Perthshire an entry reads '. . . were the two handed spinning wheel more generally used, it would probably contribute in some measure to better the circumstances of the lower class people as well as increase the material of the linen-manufacturers. There are but one or two such wheels in the parish [Auchterarder]; and it is little used in many parts of the county.'

Improvements to tools used in the manufacture of linen could be submitted to the Honourable Board of Trustees for Manufacturers in Scotland for their approval with the possibility of a financial reward. In 1780 three of these were 'improvements' to the two-handed spinning wheel, one by John Squire in Perthshire and two from Auchtermuchty in Fife. From the description recorded by the Board (Scottish Records Office NG1/20/1) we understand that Squire had moved the flyers so that they could be driven by one band instead of two, and also moved the position of the distaff nearer the spinner so that it could be adjusted to any height without restricting the size of the driving wheel.

Over the two improvements submitted from Auchtermuchty there seemed,

*According to Rettich, a spinning wheel with treadle produced 350 metres per hour and one with two flyers 498 metres per hour, if such calculations are valid when so much depends on the individual's speed of work.

according to the Minutes of the Board, to be some doubt whether or not Peter Duff had stolen the idea from Hutchinson and Drummond. However, a firm of manufacturers, Neil MacVicar, together with Andrew Nimmo, the first of a family of Edinburgh wheelwrights, were instructed by the Board to examine the wheels. In a letter to the Board they then wrote: 'from the report of the spinners who at their desire tried the wheels that they spun six cuts of yarn upon each in three hours, but were not able to do the same quantity on John Squire's wheel in less than four hours'. They came to the conclusion 'that they believed that the same, or a similar invention had already been introduced into the cotton mills, but that its application to the spinning wheel is new' (Minutes for 10 Dec. 1788). In January 1789 the Board awarded a premium of £20 to be shared equally between Peter Duff and Hutchinson and Drummond.*

Some very handsome Scottish two-handed spinning wheels still exist, often the vertical type, but Gray in his *Treatise* illustrates a horizontal type with a sloping stock and small vertical posts supporting horizontal maidens which are sufficiently long to carry two flyer mechanisms in upright bearings (**60**). Two doubled bands are here needed to drive the mechanism independently from the single wheel. It probably dates from the nineteenth century; Rettich copied this same illustration but suggested incorrectly that this Scottish arrangement was the same as that illustrated by Firmin in 1681.

A horizontal spinning wheel with a pair of parallel flyers on horizontal maidens is in the Nordiska Museum, and is of a pattern which came into Sweden in about 1840. It is signed 'jons' so may well have come from Scotland. Double flyer wheels used in Sweden and Norway, however, are not of this type, but vertical, invented in 1840 by Alois Mager, with a solid triangular base and the treadle inset into a cut-out. Though there had been mention of the use of a double flyer in Norway as early as 1779, Marta Hoffmann (1948–49) tells us that in Norway these spinning wheels were copied from the Swedish when for two years between 1854–56 an instructor from Sweden travelled round teaching the use of the double flyer. The nineteenth-century venture was not a success in either country, one of the reasons, according to Hoffmann, being that the country people had little use for the fine linen which the double flyer was specially intended for.

*There is no indication of what this 'invention' referred to, but it might be the addition of a heart-shaped cam (to be described on page 165) for winding the yarn evenly on to the bobbin without the use of hooks. However, as it is not known for how long the Scots had adopted the bobbin drag method of braking, which was used by Arkwright on his cotton machine, it could be from this period that this drive became so much used in Scotland.

Also included in the Board's list of improvements were two ingenious spinning wheels. One was: 'A wheel for spinning and by means of machinery behind for doubling and twisting yarn into thread all at one and the same time. It may be used, too, as a spinning wheel only.' Although these were considered 'a useful contrivance' by those that tried them, yet they were not considered suitable for manufactures but for 'private families who might have occasion to make a little thread'. It would seem that it was heavy work turning all this 'machinery' by one treadle since a comment on the second improvement was, 'that the motion was rather easier'.

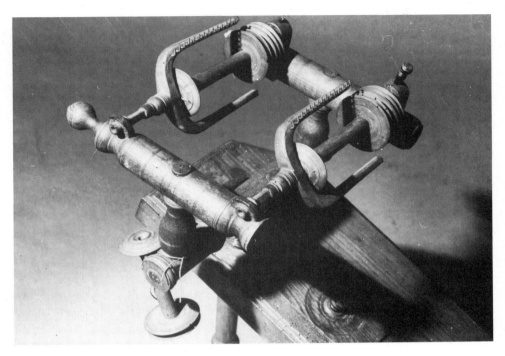

60 Detail of Scottish horizontal two-handed
spinning wheel showing two flyers side by side
on horizontal maidens.
*Photograph: National Museum of Antiquities
of Scotland, Edinburgh*

The remarkable double flyer in the Nordiska Museum, mentioned earlier,
came from the Chief Commandant of the King's army from a county north-west
of Stockholm. The treadle is made as just described; the frame of four splayed
posts supports a central cross-bar with a flyer to either side of it, and the distaff
is in front of the orifices. Two uprights from the same bar support the driving
wheel (diameter 13½in., 34cm.) which is positioned directly above the flyer
so that the footman is remarkably long. Its overall height is 51in. (130cm.) and
it is painted brown. This elongated spinning wheel is thus a double flyer 'castle'
wheel.

In Denmark, Erna Lorenzen notes that a number of spinning schools set up at
the end of the eighteenth century, several of them in Copenhagen, are known to
have used the double flyer. These too were the vertical type and not unlike
those from Sweden.

A drawing by Gabriel de Saint-Aubin, a mid-eighteenth-century French
artist who depicted daily life, shows a room with two lines of women spinning
facing two lines of children, all using the French horizontal type of spinning
wheel with frame, and with a flyer mechanism on either side of the driving

wheel, each with a tensioner at the end; the distaff holding the flax is in the centre opposite the wheel and treadle (the drawing is reproduced in *Ciba Review* no. 28). It is, according to the artist, an attempt to teach children to be ambidextrous, an idea which, he says, came from the English.

Dr Hoffmann states that in Germany a vicar's daughter circa 1750 invented a double wheel with only one driving band for both flyers. A beautiful nineteenth-century example of this from southern Germany is shown in a photograph in the archives of the Deutsches Museum, Munich; it is an upright wheel with an individual drag on each flyer and a single band across the two bobbin whorls; therefore only one tensioner is required (61). An engraving from Ulm in 1851 shows a horizontal spinning wheel with a sloping stock (slotted for the wheel) and two flyers arranged directly above each other. The maidens are therefore very tall and the back one also holds the distaff arm; there are two doubled driving bands. A fine example of a vertical two-handed wheel from Upper Austria made in about 1900 and now in the Science Museum, London, also has the two doubled driving bands; its wheel has four grooves in the rim.

The two-handed spinning wheels in the museums of Lower Saxony are very like those in Holland: horizontal type with slightly sloping stock, two mother-of-alls, two sets of maidens, and two flyer mechanisms slightly staggered on the stock to allow the two doubled bands to reach the driving wheel. Each mother-of-all is independently tensioned. But as well as these, some examples from Brunswick are very small dainty vertical types similar to the single flyer type of the region.

In Switzerland a spinning wheel from Lumbrein, in the Rätisches Museum at Chur, has an unusual structure of two pairs of diagonally crossed posts mounted upon a rectangular base; the wheel axle is supported where they cross, and the upper ends of each pair of posts are joined to hold a tensioner with each flyer mechanism just below. The distaff is in the centre.

Horner obtained a double flyer spinning wheel from the Austrian maker Josef Auinger. It is a vertical type with frame and with a flyer on either side of the tensioner. The treadle is outside the frame and the flax is arranged on a free-standing distaff. Even so it would seem awkward to draft fibres with both hands across the body when the right hand is working from further away; it would seem much more comfortable to draw down, as is the arrangement on most vertical types with base. Horner says that at that time (1904) Auinger was making 30 new spinning wheels every winter, 'two thirds with two spindles for finer yarn and one third with one spindle for coarser yarns' – a similar observation on fineness to that made by Hoffmann (page 152). Insofar as technique is concerned, one would have expected it to be the other way round, the single spindle for the finer counts, but it must have been the custom to use only the very finest quality flax to facilitate spinning two threads at once. However this does not seem to have been the case in England: a Plymouth half-penny trade token of 1795–96 portrays on one side a weaver and on the other a woman using a

61 Nineteenth-century double flyer vertical spinning wheel from South Germany. *Photograph: Deutsches Museum, Munich*

vertical two-handed spinning wheel. Around the edge of the token is written 'Sail Canvas Manufactory', which shows without doubt that the spinner is making a coarse yarn. Plymouth would seem an obvious place to find such an industry, yet according to William Burt, writing later in 1814, the town had at that time two such factories working by then with machine-spun yarn; but their proprietors 'had to contend with, at first, a prejudice among the sail-makers in the Port against the Plymouth canvas, owing perhaps in some degree to its being many years ago, an imperfect article and more resembling Scotch cloth than its present excellent quality. . . . The first canvas manufactory established in Plymouth was in Westwell Street, by the late Mr Jardine, who introduced it from Scotland, of which country he was I believe a native.' One therefore might speculate that Jardine had introduced the two-handed spinning wheel from his own part of the country to speed up production, but not apparently to improve quality.

Few examples of a double flyer spinning wheel remain in England. Snowshill Manor, Gloucester City Museum and Art Gallery and the Pitt Rivers Museum, Oxford, each have one showing certain common features and they therefore could conceivably be of this area. The one in Oxford is alleged to have come from Great Tew in the county. They are vertical, with wheels approximately 18in.

62 Scottish vertical spinning wheel with three flyers.

63 A multiple spinning wheel with four wheels and eight flyers to be used by four spinners simultaneously. c.1818.
Photograph: National Museum of Ireland, Dublin

(45.5cm.) with thick heavy rims, a single driving band and drags on the flyers.

One so often finds double flyer spinning wheels with one flyer missing. The probable reason for this is that it was removed when the pressure on spinning was not so great in regard to quantity, and spinners could revert to using only one flyer if spinning for their own household needs.

In America the two-handed spinning wheels seem to have developed a style of their own. They are the vertical type with a base supported by four legs. Above the wheel there is, in effect, a platform, to hold the two flyer mechanisms, formed of two wooden pieces joined at each end by small cross-bars, and leaving a space in which the upper circumference of the wheel rotates; this avoids raising the orifices too high. Also known in America is the double flyer castle wheel, but though in construction it does not resemble the Swedish one described earlier it is closer to the Scottish castle wheel with a single flyer. It is suggested that two handed wheels were used in America by two people, for there they have names such as the 'gossip' wheel, 'lovers' wheel or 'mother and daughter' wheel. Spinning wheels for two people have indeed been made: there is one in the collection of the Nordiska Museum, brought from abroad. The two spinners sit opposite each other, however, and it is in fact two spinning wheels joined together by a strong central bar which holds the wheel supports, each tilting slightly away from each other; there is no necessity for the spinners to synchronise.

In Scotland an occasional spinning wheel with three flyers is seen. A good example is in Old Leanach Cottage, Culloden (62). The three flyer mechanisms are mounted on a single front horizontal maiden, the middle one of the three being raised higher than the other two, all with independent tensioners within which the ends of the spindles rest, in wood chocks. The orifices all would appear to have the same width. A similar arrangement (now with the central flyer missing) is on a spinning wheel preserved in the National Museum of Antiquities of Scotland, Edinburgh, made by Jas. Clement, Crieff. It is said to be a two-handed flax wheel which will convert to a single-handed wool wheel (for this using the centre flyer). It would also be possible to use the centre flyer for plying.

But perhaps the most amazing multiple spinning wheel is in the collection of the National Museum of Ireland, Dublin, and comes from Belleek, Co Fermanagh (63). It incorporates four spinning wheels built on a single frame with four separate driving wheels each turning two flyer mechanisms placed one above the other, the lower one slightly to the right of the upper. There are four distaffs. The treadles are interconnected so that synchronisation by the four spinners is essential: imagine it in use! P. F. Nyhan suggested that this 'machine' may be a development of a design by Peter Bernard, Inspector of the Linen Board, 1819. In Ireland, as in most other countries except perhaps Scotland, the double flyer, let alone the eight-handed one, was never adopted in a big way. McCall recalls a plan for spinning with both hands, introduced at the beginning of the nineteenth century in Lisburn (which shows perhaps how little known were the schools at

Dundalk and Greenvale further south) but it was too late for it to be of any consequence and the scheme did not last very long.

BOUDOIR OR DRAWING-ROOM SPINNING WHEELS

In the Bayerisches Nationalmuseum in Munich there are six little spinning wheels with a floor frame, long spindly legs and a little platform with a gallery round it. One has a drawer underneath. Four short legs support the horizontal frame, which carries the flyer mechanism on the left and two short uprights support the tiny wheel, about 7in. (18cm.) in diameter, on the right. As would be expected from Bavaria the drag is on the flyer. They are beautifully carved, delicately and extravagantly; one in particular has a treadle carved in the shape of a horse's head facing upwards, with the lower end of the footman pinned into its mouth.

When in the second half of the eighteenth century it became fashionable for ladies to spin, they did so as a way of passing an hour or two in the morning; having donned voluminous aprons to cover their gorgeous clothes, they did a little embroidery, sewing or flax spinning. The tall elegant distaffs were no doubt dressed by a maid. Such little wheels were ideal for these ladies and became known as boudoir, drawing-room or toy spinning wheels. They must have been quite popular in England since there are many of these dainty little objects in our museum collections, and well-preserved specimens are highly valued as antiques. More often than not the wheels are made of lead or the rims are lined with lead, brass or pewter to give them the necessary weight they otherwise lacked through being so small. Unfortunately this could make them too heavy for their delicate supports and many of these are now found to be broken. They usually have both a handle on the front axle and the long footman attached to the treadle at the back.

Some of these boudoir spinning wheels are referred to as carriage wheels. These may have four little pegs which fit into the edge of the table and through the top of the tall legs enabling the spinning wheel to be taken apart. In this case the footman is attached to the front axle, so that when the legs are removed it is unhooked, and an ivory or wooden knob can be screwed on in its place to form a handle. In this way the spinning wheel could be used in a carriage if so required. A pretty example made of mahogany and fruitwood, in the Farnham Museum, has a single vertical wooden screw inserted from below in the centre of the neat rectangular stock, so that the platform with the legs attached can be removed, leaving a table spinning wheel standing on four ball feet. With its flyer drag, leather driving band, leaded wheel rim, ornamental wheel points, bone or ivory decoration and the woodwork well turned and carved, it is probably from Bavaria or Austria dating from the middle of the eighteenth century.

Boudoir wheels were usually made of expensive, beautiful woods and some are lavishly turned and decorated with marquetry or inlay. Others are painted or lacquered, often cream-coloured or pale yellow and decorated with shell, fish or feather motifs. A fine English example in the Victoria and Albert Museum, exhibited in the costume department, has a cage distaff under which there is a ring attached to the post. It may therefore have been used for plying, the skeins placed on the distaff and the threads kept close together and guided to the orifice by passing through the ring. (The *Encyclopédie* states that there must be a guide ring for plying.)*

To what extent these boudoir spinning wheels were directly inspired by the little wheels of Bavaria is open to speculation, but certainly they were also found in other fashionable places in Europe besides England. It is also not clear how far they were imported or made here, but John Jameson, owner of a toy and turnery manufactory in Goodramgate, York, advertised in the *York Courant* on the 22 August 1780 'all kinds of toys and pieces of useful invention'. His advertisement continued: 'It is remarked by travellers in the City of Nuremberg in Germany that there are neither beggars nor poor rates owing as supposed to the immense quantities of toys and small wares manufactured there by the inferior class of people. J. Jameson therefore intends to employ the industrious poor in this manufactory and hopes to have the encouragement of the public in the undertaking.' On the 17 November 1789 (by which time he had moved his shop to Little Alice Lane, College Street) he advertised 'a large assortment of curious and useful articles' including 'Best inlaid and mahogany German and other spinning wheels'. It would therefore seem likely that as well as having travelled in Germany, he was making spinning wheels of German types in his own workshops. When on 2nd August 1802 he disposed of his stock-in-trade, cabinet-maker, turner and toyman John Jameson advertised a 'great variety of spinning wheels'. A further advertisement quoted by E. W. Gregory (1927) states that 'for 20 years he [John Jameson] has supplied the first families in England and Scotland particularly in the articles of German and other spinning wheels'. A spinning wheel in the Pitt Rivers Museum collection has JOHN JAMESON carved on the inside rim of the drawer. It has a doubled band drive, unlike those in Bavaria, which provides some evidence that it was probably made here (64). Jameson was by no means the only maker of toy spinning wheels, and another northern maker who signed such wheels was T. Brown of Highgate, Newcastle-upon-Tyne.

From Sir Ambrose Heal's collection of Trade Cards we learn of one John Alexander, Ivory and Hard-wood Turner, who had his business near the Monument, London. He is listed in the directories between 1777 and 1793 and included spinning wheels amongst a hundred other things. Such makers may not particularly have made this type of boudoir spinning wheel since small but elegant horizontal spinning wheels with sloping stocks were also fashionable.

A spinning wheel in the collection of the National Museum of Antiquities of

*An example of a boudoir spinning wheel in the Royal Pavilion, Brighton, is in Mrs Fitzherbert's drawing-room but, alas, it has only been there a few years and is no indication that the wheel either belonged to her or that she could spin.

64 'Boudoir' spinning wheel, late eighteenth-century, made by John Jameson of York; wheel diameter 7¾in. (20cm.).
Photograph: Pitt Rivers Museum, University of Oxford

Scotland shows a definite attempt to copy the style of the boudoir wheel; it also incorporates the fashion for articles from the East, the wood being turned in imitation of bamboo. The wheel is decoratively painted on one side only, to please the spinner. In front of the gallery is a small silver plate stating that the maker was Willm Cunningham/Stirling.

It must not, however, be assumed that the ladies using these decorative spinning wheels were never good spinners. McCall writes: 'The first Lady Londonderry prided herself very much on her skill as a spinner; and during her periodical sojourns at Mountsteward the peeress frequently spun yarn so fine as twenty-five hanks to the pound. The very handsome toy wheel used by her lady-ship was among the curiosities exhibited at the museum of flaxen products. . . .'

TABLE SPINNING WHEELS

Another form of drawing-room spinning wheel, again with a small driving wheel, is the table model which could stand decoratively, but out of the way, on top of a piece of furniture. It is, of course, turned by hand (like the original flyer wheels) with a handle on the wheel axle to the spinner's right; most examples have a distaff attached to the base on the left. They were only suited for one-handed spinning of flax, but like the boudoir type were probably used as much for plying silks for embroidery or linen for sewing thread as actual spinning. Those without distaffs were most likely only for twisting, particularly where the orifice is not too small, although silk waste or even cotton, first prepared into a roving, could have been spun on them.

A Meissen porcelain figure of about 1750 portrays a lady with a spinning wheel (without distaff, but later models have one) placed on the table beside her. It was designed by Johann Joachim Kaendler from an engraving by Louis Surugue (1686–1762) after a painting by Jean-Baptist Simeon Chardin. There is such a statuette in the Victoria and Albert Museum, where there is also a beautiful example of a table spinning wheel. It is enriched with marquetry of tortoiseshell, mother-of-pearl and ivory. The rim of the wooden wheel is decorated with running animal motifs and the wheel is turned by a knob embellished with ivory attached to a curved crank. The flyer is of brass and the mother-of-all of solid ivory, while the back maiden is extended to support an exquisite little water pot. The distaff is topped with a silver ring, not unlike the one illustrated in the *Encyclopédie* (see page 126). It is pointed out by Peter Thornton ('Italian Crafts-manship in Wood') that the base, reminiscent of the work of Pietro Piffetti, working in Turin in the middle of the eighteenth century, is of superior work-manship to the spinning components and that these were therefore probably made in a different workshop. Another exquisite table spinning wheel, only $9\frac{1}{2}$in. (24cm.) high, owned by an antique dealer in Munich, is believed to be from either northern Italy or Austria, and made in the first half of the eighteenth century.

The main problem with these little spinning wheels is how to keep them steady while working with them. An engraving from the second half of the eighteenth century shows a lady using one balanced across her knees. She sits spinning short flax which is folded over and across the top of the distaff and tied in place with all the ends pointing downwards, from which she draws with her left hand while turning the wheel with her right. Her companion is engaged in embroidery.

Not all table models were so extravagantly designed as the one in the Victoria and Albert Museum, and are more often made of wood with the rim of the wheel weighted with brass or lead. Wheel diameters vary from 7in. to 11in. (18–28cm.). A. Hamilton of Edinburgh overcame the problem of keeping a table model steady by setting a number of lead weights into the underside of the base (65). It is quite usual to find two little brass rings attached to the side of the base, by which some form of clamp could secure the spinning wheel to a table top.

Preserved in the house of Beatrix Potter, the writer of children's books, who did not herself spin but liked objects which she felt suitable for her home, is a table model. It has a tapered base 13in. (33cm.) long and a wheel diameter of 11½in. (29cm.) and a handle; not only are there two little brass eyes, one at either end of the base, but there is a hole lined with brass in the centre of it, which makes one wonder if it had once been attached to a stand or even hitched up to a treadle (as is the case with the one mentioned on page 158 in the Farnham Museum).

Table spinning wheels can, however, easily be confused with a type of lace maker's bobbin winder. The feature of this is a little wooden holder into which

65 Table spinning wheel made by A. Hamilton, Edinburgh; wheel diameter 10½in. (26·5 cm.). *Photograph: National Museum of Antiquities of Scotland, Edinburgh*

the bobbin was packed, in the place where the flyer is on a spinning wheel. Attached to the base of the bobbin winder is an articulated arm supporting a four-armed skein holder.

GIRDLE OR BELT SPINNING WHEELS

Girdle or belt spinning wheels were another type made to amuse ladies in the second part of the eighteenth century, and very charming little toys they were too. A particularly well-preserved example in working order is in the Science Museum, London (**66**). Its wooden holder is designed with two horns so that it can be slipped into a belt round the spinner's waist. From the centre of this holder and at right angles to it, a short turned wooden arm holds the entire spinning mechanism. This consists of two cog wheels encased in brass, the larger of the two being turned by a small handle on the right. It is turned clockwise, the larger cog engaging with the smaller to turn the flyer in an anticlockwise direction for S twisting. The flyer, also of brass, has cut-out holes and slits suitable only for a very fine thread. The little lantern distaff is attached on the left-hand side, and can be positioned by loosening a brass wing-nut, similar to the nut for tightening the friction band, this being round the bobbin. Other examples of girdle spinning wheels are in the Hall i'th' Wood Museum, Bolton, and the Pitt Rivers Museum, Oxford. All three are engraved on the holder 'James Webster, Clock Maker, Salop'. (It is not unusual to find clockmakers interested in spinning wheel mechanism, and in the late eighteenth century one or two made spinning wheels themselves for the ladies.)

The Websters were a family of clockmakers: from an article in the *Shrewsbury Circular* in December 1940, written by a local historian, L. C. Lloyd, we learn that James appears to have been the first to enter the clock business, for certain by 1745. His third son, Robert was born in 1755 and too entered the business; it is not certain when. Robert was the most inventive of the family, having patented a washing machine and also a mangle.

In 1791 there appears the following entry in the *Shrewsbury Chronicle*: 'On Saturday the 12th inst. three most beautiful and elegant spinning wheels, viz. a foot-wheel, a table-wheel, and a girdle-wheel, were presented to the Queen at Windsor, which were most graciously accepted and highly approved of by her Majesty. The two first wheels are the entire invention of Mr. Robert Webster, Clock and Watchmaker of this town; the mechanism is such, and the motion so fine as not to emit the least noise or sound, so that they may with the greatest propriety be called Dumb Wheels.' Although we see from this extract that Robert Webster actually 'invented' two of the spinning wheels there is no claim that the girdle wheel was his. As the only existing examples in this country all have the name of James Webster, so it may have been one of his father's spinning wheels that Robert also took to Windsor.*

The drawing of Mrs Lane actually using a girdle spinning wheel, for what appears to be plying from two little skeins of thread, dates from before the presentation at Windsor Castle (Frontispiece).

In Germany the girdle wheel is called a *Damenspinnrad* (lady's spinning wheel) and a copper engraving of 1785 shows three ladies sitting round a table drinking coffee each 'wearing' one and spinning flax from a tiny distaff firmly tied round with a *Kunkelband*. The ladies are all richly dressed (one is wearing a large hat trimmed with feathers) and all draw the flax with their left hand and turn the little handle with their right. One of them, at the same time, is engrossed in a book.

---

*In the collection of Snowshill Manor there is a spinning wheel with the wheel beneath a semi-circular mahogany table which is marked INVENTER (*sic*) R. WEBSTER SALOP on the metal plaque in plain letters, unlike the flowery lettering used by James. Much of the mechanism is missing and it is now more suited for winding bobbins than spinning, but it may be similar to one of the two types invented by Robert Webster which he took to Windsor.

66 Girdle spinning wheel made by James Webster, clock maker, of Mardol, Shrewsbury. *Photograph: Crown Copyright, Science Museum, London*

The Deutsches Museum, Munich, had an example of a girdle wheel; this was stolen, but a photograph shows that it was very similar to those just described except that it had only a single horn to tuck into the belt and the distaff was a miniature tow distaff; it was dated to around 1780.

In the Nordiska Museum a girdle spinning wheel, unsigned and undated, has a two-horned holder made of fruitwood and delicately turned; the gears are enclosed in a tortoiseshell casing edged with brass and the bobbin length is $2\frac{1}{8}$in. (5·4cm.), exactly the same as on the one made by James Webster previously mentioned as being in the Science Museum. The distaff of the Swedish one is of a solid, slightly conical construction around which some flax remains with a *Kunkelband* and ribbon to keep it in place. It was purchased by the Museum in 1876 and could equally well have been inspired by the English version as by the German, or even made in one of these two countries. In both Germany and Sweden a spinning wheel incorporating a musical box was another diversion for the ladies.

AN IMPROVED SPINNING WHEEL

At the same time that inventors were at work improving the efficiency of spinning machines there were others who were still devoting energy to improving the flyer spinning wheel by devising a mechanism for winding the thread evenly on to the bobbin. To achieve this there are two alternatives: either the bobbin must move to and fro or else the flyer.

An entry in the *Transactions for the Society of Arts* for 1793 states: 'Twenty guineas were this year voted to Mr. John Antis, of Fulneck, near Leeds, for his ingenious method of causing the bobbin of the common spinning-wheel to move backward and forward; by which means, the time lost by stopping the wheel to shift the thread from one staple, on the flyer, to another, as has hitherto constantly been practised, is avoided. . . .'

The basic idea was to have an oscillating arm which moved the bobbin backwards and forwards along a metal bar (the spindle) of a little more than twice the bobbin's length. The arm was driven from an extension of the wheel axle engaging with a small toothed wheel at right-angles to it, built into the front upright, above which a circular piece of wire pushed the oscillating arm by means of a pin on it. A weight tied to the metal bar passed over a pulley, rising and falling as the bobbin advanced and receded in order to keep the pin in constant contact with the wire. It was then only necessary to provide a single ring or eye at the end of one of the flyer arms to guide the thread on to the bobbin.

In a letter to Mr More, Secretary of the Society of Arts, Antis says that his improvements could easily be adapted to old wheels and would add only a 'mere trifle of expense to a new wheel'. He continues: 'I had it tried by a lady here, who sometimes spins for her diversion, who was much pleased with the invention, and thought it might save a person at least two hours, if not more, in a day; which would be a great object for poor people.' In a second letter he writes: 'The invention has also this advantage that, whereas, at best, the old method always winds in ridges, if a thread breaks, by reeling the yarn, one may as well throw the whole bobbin away, as the thread cannot easily be found again: but this always winds across upon one another; by which means the thread can never be lost.'

However this improvement was not entirely satisfactory and in 1795 John Antis was awarded a bounty of 15 guineas 'for this further improvement to the common spinning wheel'. In an abstract of a letter he points out the defects and how he had improved them. '1st: Besides the desagreeable catch which the pinion with one leaf gave at every revolution of the wheel, the bobbin moved only by sudden jerks, and did not lay the thread on it so even as might be wished: to remedy this, I adopted the motion of an endless screw, working a toothed wheel with a heart-shaped piece of brass [a cam] fixed thereon. 2nd: As the spinner must be always perfectly at liberty to hold the thread at pleasure, and not to let it in until it is sufficiently twisted, I observed, that as the bobbin moved upon a square, the holding of the thread so affected it, that particularly when the bobbin began to be filled, and consequently the purchase further from the centre it became quite stiff upon the square and incapable of moving, notwithstanding the action of the weight: and afterwards, when the thread was left at liberty, it would jump at once half an inch or more . . .' His solution to this problem was to exchange the doubled driving band for a single band on the flyer which 'therefore needs no screw to tighten it; for, should it, even in time,

67 Detail of spinning wheel by Doughty of York, c.1795–1800. Made of mahogany with brass wheel, diameter 12in. (30·5cm.). It shows the mechanism for automatically winding the yarn evenly on to the bobbin.
Photograph: Castle Museum, York

become so slack as not to do its duty, it may easily be tied or stitched a little shorter'.

With these further improvements the mechanism was the same as shown in plate 67. The heart-shaped cam is on the opposite side of the upright and can be seen engaging with a roller, which took the place of the pin on the oscillating arm. This arm divides in two at the point where it passes through the slot in the cross-bar and 'pinches' on to a metal neck extending from the bobbin. A screw with an ivory head (just visible above the slot) controls the amount of 'pinch' on the bobbin and so replaced the need for a friction band.

Andrew Gray in his *Treatise* illustrates a vertical spinning wheel with a similar type of mechanism which he calls Mr Spence's improved spinning wheel. Rettich illustrates the same wheel, saying that the Englishman Antis first put his name to this improvement and that the wheel was constructed by the Englishman Spence. From Gray's drawings it can be seen that the mechanism is not constructionally exactly the same as that of Antis, and that he used a doubled band. It would also appear that Spence's mechanism was made at a later date since Gray, writing in 1819, says: 'As every alteration in a machine cannot be termed an improvement, therefore this alteration on the spinning-wheel being of a recent date, we cannot determine whether it is an improvement or not until its merits be fully proved by experience.' Surely 20-odd years would have been long enough to experience this 'new' improvement. Anyone who has spun on a wheel with one of these mechanisms will have appreciated its very smooth and regular action, so suitable for the rhythm of flax spinning.

More than one Yorkshire spinning wheel maker used Antis' type of mechanism. In the Castle Museum, York, there are two vertical spinning wheels which incorporate it, marked 'Doughty of York' on a circle of ivory round the orifice. The woodwork is mahogany, and the wheel has a lead rim and six squared spokes. The bobbin is 2in. (5cm.) in length and the spindle $5\frac{1}{4}$in. (13·4cm.), giving ample room for the full traverse of the bobbin. The brass flyer has one eye at the top of one arm to guide the yarn and a second eye on the curve to prevent it from touching the bobbin. The orifice is $\frac{1}{8}$in. (3mm.), and there is an articulated arm for a distaff and another with a recess for a water pot. The curved triangular base gives plenty of space for the spinner's knee.

Marshall and Doughty were toy, turnery, umbrella and cabinet manufacturers, with a shop in Coney Street near Mansion House, York, where they sold all kinds of toys. In the *Yorkshire Herald*, February 1795, there appeared at the end of their advertisement: 'N.B. The new-invented spinning wheel the most complete ever offered to the Public which winds the thread on the pearls in a cylindrical manner, and prevents the Ladies having the trouble of altering the thread on the Feather.' Joseph Doughty himself died in December 1801, but his widow, Martha, announced in the *York Courant* the following January that she would continue the business in Coney Street. The firm was then called Marshall, for a vertical spinning wheel with two ivory discs on the front upright has these engraved 'Marshall late Doughty' on the top disc and 'York' on the other.*

While the Marshall spinning wheels were much the same to look at as those made by Doughty, though the turned wooden wheel spokes are of inferior workmanship, an improvement had been made to the mechanism. Instead of the $3\frac{1}{2}$in. (9cm.) cog wheel (as seen in plate 67) the endless screw on the wheel axle engages with a horizontal gear wheel with a central axle pin engaging with a second vertical cog wheel. The mechanism is almost totally enclosed within the front upright, with only the vertical cog and the smaller heart-shaped cam protruding; this arrangement of cogs makes the traverse of the bobbin considerably slower and therefore the yarn more closely packed on the bobbin. It is an improvement mechanically, but should the spinner lose her thread, it could be more difficult to find. The spinning wheel must have been made prior to 1814 since in that year Mrs M. Marshall sold her stock to John Hardy.

*Whether the widow married Marshall or whether this was her maiden name to which she reverted after the death of her husband, adding the Mrs, I have not been able to ascertain, though I think the latter is most likely. A spinning wheel illustrated in the advertisement section of the *Apollo* magazine, August 1938, shows a vertical spinning wheel and the automatic wind-on mechanism. It bears the name 'Doughty, York' – and is dated 1807. It is not made clear whether the date was found on the spinning wheel or was an estimated date by the antique dealer advertising the wheel. If it had been actually dated, it would mean that the name Doughty continued to be used for six years after his death. My own opinion is that the antique dealer did not know the exact date of Doughty's death.

Hardy too continued to make spinning wheels of the upright type with the automatic even wind-on of which one example has an ivory disc engraved 'Hardy late Marshall' (both this and the one described above are privately owned). He also made horizontal spinning wheels on traditional lines since in the Castle Museum, York, there is a dainty little model, probably made of sycamore, stamped with the name Hardy. However, when in 1832, he advertised in the *York Gazette*, although still as a turnery business, spinning wheels were not among the list of nick-nacks that he sold.

Perhaps the best-known spinning wheel maker to use Antis's improvement is John Planta, who also lived in Fulneck. As well as making vertical spinning wheels in much the same way as Doughty, he made unique and beautiful spinning wheels and incorporated the mechanism horizontally instead of vertically; the cog wheel is encased in brass and the brass driving wheel measures approximately $11\frac{1}{2}$in. (29cm.) in diameter. The wheel axle extends forward past the cog wheel and ends in a small handle with an ivory knob for positioning the wheel before spinning commences. The oscillating arm is attached on the right and moves to and fro in front of the wheel. The mechanism rests on the top of a small platform approximately 18in. (45·5cm.) wide and 9in. (23cm.) deep, with a drawer underneath opened and closed by a spring catch and used for storing spare bobbins, all supported by four slightly splayed tapered legs. The wheel is part-concealed, and the whole has the restrained formal grace and simple lines of the Sheraton style (68); made of mahogany, some are more sumptuous than others, with inlay of satin wood or tulip wood, spokes with inset corner strings, ivory fitments and a little brass urn sitting in the centre of the diagonal stretchers, the urn being echoed in ivory at the top of the distaff. Some examples have an articulated arm attached to the platform to hold the distaff, and another for a light or a little pot for water. Some have egg-cup-shaped pots attached directly on to the platform and lined with glass, and were likely to have held a wetted sponge. There are six little holes for oiling at the necessary points in the mechanism.

These spinning wheels are now more valued as pieces of antique furniture than for their spinning capabilities; it would seem that they were made between 1798 and 1802. Some, but not all, have a label inside at the back of the drawer saying: 'Made by John Planta, at Fulneck near Leeds'. Planta was always discreet over his labelling and on the vertical ones the label is underneath the base. This may be because he worked within the community of the Moravian Church. Christopher Gilbert (1970), in his article about John Planta, tells us that he was born in Jamaica in 1764 and came to the Moravian Church of United Brethren in Fulneck in 1798. The Moravians, a Christian sect of German origin, formed a self-sufficient community, a principal occupation being spinning and weaving, but other trades were also followed, including clock-making and cabinet-making. Planta therefore would have been well conversant with the workings of a spinning wheel and with his gift for cabinet-making, and by using Antis's

68 Mahogany spinning wheel in the
Sheraton style made by John Planta of Fulneck
near Leeds, c.1800. Wheel diameter 11¼in.
(28·5cm.). Now in the Science Museum,
London.
*Photograph: Crown Copyright, Victoria and
Albert Museum, London*

improvement, he was able to produce a wheel which appealed to the gentry for their drawing-rooms. However, from the Sisters' House Cash and Account Books it seems that after 1807 he did humbler work which included furniture repairs, miscellaneous jobbing, and that he made only various small articles up to his death in 1824.

A number of these spinning wheels are still in existence: the Science Museum (two); Heaton Hall, Prestwich; Cannon Hall, near Barnsley; Temple Newsam House, near Leeds, as well as in private collections. There is also an example in the Wythe House, Colonial Williamsburg, Virginia, purchased in 1959. Not all are in working order and the one at Temple Newsam House has unfortunately been altered to become a winder. Not only is there no longer a flyer, but the front maiden has been replaced by a mahogany upright without an orifice, with a spring catch arrangement to release the bar carrying the bobbin. It was probably made originally for Mrs William Rhodes, mother-in-law of Benjamin Gott, the first of the great Yorkshire wool spinners. It is known to have started its life as a spinning wheel since a member of the Gott family remembers being shown towels made from the flax spun on it.

In common with the Doughty vertical spinning wheels, those made by Planta have their orifices over 30in. (76cm.) off the ground; this is unusually high, and for comfortable spinning would require a chair at least 19in. (48cm.) high. One such chair, with a cane seat and narrow depth of 9½in. (24cm.) is in the collection of Snowshill Manor and would have been suitable for use with such spinning wheels. There is some doubt as to the extent to which these narrow chairs were in fact used for spinning, particularly as pictures of spinners do not seem to show them. However, some bone models of ladies spinning made by French prisoners about 1810 show them perched on straight-backed narrow-depth chairs.

QUEEN VICTORIA'S SPINNING WHEELS

As the fashionable pursuit of spinning rose to a hey-day during George III's reign, we hear of his consort's interest in it. Mrs Delawny, a lady who enjoyed royal society, wrote to her friend Mrs Frances Hamilton in 1781: 'The Queen etc. came about twelve o'clock and caught me at my spinning wheel (the work I am now reduced to) and made me spin on and give her a lesson afterwards; and I must say did it tolerably well for a Queen. She staid till three o'clock; and now I suppose our royal visits are over for this year.'*

*Mrs Delawny had excelled at embroidery and shell work, but at the age of 81 her sight was failing so she had turned to the occupation of spinning. Thomas Firmin remarked that older people could still spin a tolerable thread even if their eyes were not good, while McCall points out, 'Besides the mental power called into exercise by hand-spinning, the continued action of finger and thumb in twisting the flax during the process of production imparted the most wonderful degree of sensitiveness to those members, and this delicacy of touch became intensified in proportion to the fineness of the spinning. Many of the most successful spinners of ordinary yarns in Northern Ireland were blind women. In these instances, the loss of one sense had so powerfully influenced the quickness of perception, as to give them vastly increased powers in the others.'

Unfortunately, the whereabouts of Queen Charlotte's spinning wheels is not known but, as well as receiving them as gifts, the Queen also gave them, as can be seen by these words of McCall: 'Mrs Turtel, wife of an extensive farmer at Aughagallon, near Lurgan, was one of the most celebrated spinners of that district of Antrim. This lady spun, from flax grown in her neighbourhood, a parcel of yarn so exceedingly fine that three hanks of it could have been passed through a wedding-ring. That yarn was afterwards woven in a cambric web by a superior weaver, and the cloth, in its brown state, was quite a marvel of fineness and beauty. When bleached and ornamented, the web was forwarded to Windsor Castle as a present to Queen Charlotte. Her Majesty expressed herself highly gratified . . . and in return forwarded a handsome present to the donor, . . . an exquisitely-constructed spinning-wheel, formed of mahogany, together with a fancy reel of the same material.'

It appears that the Queen sent to Scotland to have spinning wheels made, since George Nimmo, one of the family of Edinburgh wheelwrights already mentioned (page 152), is listed in the Edinburgh Trade Directories of 1820 as 'Wheelwright to her late Majesty'. Queen Charlotte died in 1818.

However, it was not as a fashionable pursuit that Queen Victoria learnt to spin but through her involvement with the Scottish way of life. Frank Pope Humphrey writing about the Queen at Balmoral says: 'I never heard that she took particularly to needlework of any kind – that refuge of Queens in former ages from dreary monotony of their lives – though occasionally she has sent to some lucky bazaar a specimen of her knitting. Lucky, for her work fetches nearly or quite its weight in gold. And she once learnt to spin upon a little wheel, most artistic and graceful of industries. One of her cottage women at Balmoral taught her to spin and she spun enough for a napkin or two. An old cottager told me she had once several threads of the Queen's spinning, but that she had given them away to eager petitioners.'

In the 1868 Inventory for the Queen's sitting room at Windsor Castle two spinning wheels were listed; one, an ebonised spinning wheel without a maker's name, and the other a mahogany spinning wheel made by Peter Stewart, Spital-field, Dunkeld. In the 1860s the Queen was photographed sitting at a spinning wheel which could have been the first of these two, since the wheel looks dark and shiny, the result of the ebonising which was a finish now and then used on spinning wheels. The wheel, simply turned, is of a sturdy design, horizontal with a sloping stock, and the footman is just a length of cord. The distaff upright is on the left back corner of the stock. She used both hands to spin the flax, the left drawing the fibres.

At Osborne House there is a spinning wheel of light varnished wood, perhaps beech, inscribed 'To The Queen from the Duchess Dowager of Athol, 1866, made by John McGlashan, aged 70, Easingeal, Blair Athol'. This is a simply-designed horizontal treadle spinning wheel with a fairly steep slope. The wheel, of oak, diameter 16in. (40·5cm.) has a handle on the axle. It is interesting to find this tradition still continuing into the second half of the nineteenth century. There is also a reel made of light mahogany superficially matching the spinning wheel, made by P. Stewart of Dunkeld. Unfortunately there is no conclusive evidence that the spinning wheel was kept at Osborne House during Queen Victoria's lifetime, although it was there when the Office of Works took over the house early in the present century.

Dr Norman Macleod, who became Chaplain to the Queen in Scotland in 1857 and was also her friend, wrote to his wife in 1866: 'After dinner the Queen invited me to her room, where I found the Princess Helena and the Marchioness of Ely. The Queen sat down to spin, at a nice Scotch wheel while I read Robert Burns to her. Tam O'Shanter and A Man's a Man for a' That, her favourites.' A woodcut of this scene by R. Watson shows that the Queen's spinning wheel at Balmoral was different from the ebonised one at Windsor, since the maidens, instead of being vertical, are horizontal as in some of the Scottish horizontal double flyers.

A Belfast firm, James McCreery & Son, was listed in the 1880 Belfast Trade Directory under 'Carvers, Gilders and Picture frame makers', but in 1887 advertised itself as 'Carvers in Irish bog oak, spinning wheel manufacturers, also Dickson's patent collapsible post office sorting table. Established over half a century'. There then followed a formidable list of those who patronised the firm, headed by Her Majesty the Queen, followed by the Presidents of France and of America, the Duke and Duchess of Marlborough, the Right Hon. W. E. Gladstone, the Lord Lieutenant of Ireland etc., and a list of prize medals won in Belfast, Cork, Paris and London.

An example of a McCreery spinning wheel is in Lotherton Hall, Aberford, near Leeds, Yorkshire. Made of birch, mahogany, oak and bog oak, it is a horizontal type with a carved trail of shamrocks along the edge of the stock and a treadle in the shape of an Irish harp. A bog oak cross and a lighthouse tower are inlaid into the distaff upright. The Ulster Museum has a photograph of a spinning wheel showing much the same characteristics with the bog oak cross and the harp-shaped treadle, but thistles, roses and shamrocks are carved on the edge of the stock instead of the trail of shamrocks (69). The bows of wide ribbon tied round the flax and mother-of-all make one think that the photograph was taken at one of the medal-winning occasions, perhaps in Belfast in 1870 or 1876. Although unsigned, it is known that the example at Lotherton Hall was made by McCreery, since when it was acquired it was accompanied by his trade card stuck on to the back of a photograph of the elderly Queen Victoria sitting at a spinning wheel (70). This is not the same spinning wheel that Her Majesty was using in the photograph taken in the 1860s, nor was this one made by McCreery, since he made his maidens vertical, but it is similar to Watson's woodcut with the horizontal maidens. It is also possible to see in the photograph that there is a bearing on the right of the flyer which could take a second flyer (also just visible is a piece of ribbon tied to the end of the maiden, undoubtedly attached to the

69 Spinning wheel by James McCreery of Belfast; second half of nineteenth century. *Photograph: Ulster Museum, Belfast*

threading hook). I have seen a photograph of a spinning wheel with these same features, now in private hands but alleged to have been a present from Queen Victoria to a friend at Windsor; it has a metal plate which reads 'Peter Stewart, Spitalfield, Dunkeld'. It would seem, therefore, that McCreery used the photograph of the Queen merely for advertising purposes. Whether he claimed her patronage because she actually bought spinning wheels from him is difficult to ascertain. Marion Channing (1969) writes of a photograph she has seen of the Queen sitting at a spinning wheel which was presented to her by McCreery of Belfast. But Channing's description of this wheel as a 'regulation style Saxony wheel' does not match with the ornately carved spinning wheels which we know McCreery to have made and which one would have expected him to have presented to the Queen. It is possible that the Queen, with her ability to spin, preferred a plain serviceable spinning wheel to a decorative one and that the photograph shows another design made by McCreery; it could even be the unsigned spinning wheel that was in the sitting room at Windsor. It would seem however, that it was Peter Stewart, the Scottish maker, who received most of the royal patronage.

Another Scottish spinning wheel maker, Hugh McMaster of Kilpatrick on the Isle of Mull, used the Queen's name and labelled his wheels 'The Lady Victoria'. This was perhaps to promote the sale of his wheels late in the century when the Queen had returned to popularity and the more so if it were known that she could spin.

Queen Alexandra could also spin. We know from a newspaper cutting of 1925 that she was taught to spin flax at Sandringham, since her teacher Mrs Elizabeth Pepper from the Lake District was presented with a vertical spinning wheel by the Queen in appreciation.

70 Queen Victoria at her spinning wheel. The reverse side of McCreery's trade card.
Photograph: The Leeds City Art Galleries

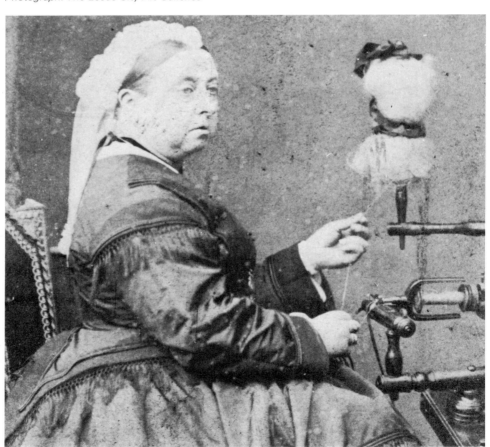

Chapter Seven

Spinners and spinning

When Adam delv'd and Eve span,
Who was then a gentleman?

Spinning has always been connected with women, and the spinning wheel, once established, became a symbol of virtue and thrift, for indeed the industrious housewife spent every spare moment spinning. However, in folk and fairy tales as many relate to the lazy spinner as they do to the good housewife; in Grimm's story *The Hurds*, it is the lazy mistress who would only spin the best-quality flax and threw away the hurds (tow), enabling her servant girl to pick them up, spin them and so win a husband for herself. In all Grimm's stories the fibre is flax; one of the tasks set to the captured girl in *The Water Nixie* was to spin tangled dirty flax. And surely, the fact that the fibre increases in value once it is thread, forms the basis of the story of *Rumpelstiltskin*, who for three nights in succession spun straw into gold for the miller's daughter, who thus became a queen.

'Spinning wheels have also been incorporated into fairy lore' writes I. F. Grant (*Highland Folk Ways*) 'and it was considered inadvisable to leave the band on the wheel when the household retired for the night for fear that the fairies might then use the wheel'. This may be a cautionary tale since anyone tampering with the spinning wheel could entangle the thread or lose the end. The wise spinner winds the last few inches of spun yarn round the top of the front maiden, which prevents the yarn from untwisting or getting lost.

In Britain, where wool has been of so great importance, a folk belief in Scotland, also in the Lake District and the Isle of Man, is that the wool spins better when the sheep are at rest. Wordsworth used this theme in his *Song of the Spinning Wheel* and it is undoubtedly a piece of folklore based on fact, since the warmth from a fire eases the natural grease in the wool and facilitates teasing and spinning. Long ago this beneficial effect may not have been consciously realised when the family gathered round the hearth in the evening and the women would spin.

If there were get-togethers, the women frequently took their spinning wheels along and spinning parties were a common occurrence, particularly in country areas, not only for gossip but there would be jollification and singing. With so many young girls at their spinning wheels many of the spinning songs were about love and their dreams rather than the job in hand, though perhaps the most disliked job was carding.

It is interesting to note that while in the first place the word spinster referred to someone who spun, in the seventeenth century it was the legal term for an unmarried woman, and therefore often someone young. It was only later that it became a word for a woman beyond the usual age of marriage, an old maid.

At the end of Christmas Festivities, the day after Twelfth Night, January 7th, was known as St Distaff's Day for which Herrick wrote a verse:

> 'Partly work and partly play
> You must on St Distaff's Day. . .'*

In most country households spinning was a job for all the women in the evening after the day's work was over, but some sharp employers stipulated that servant girls should be at their spinning wheels by six in the morning. *The Statistical Accounts of Scotland* (1792) for Kirkden, Angus, tells how many girls 'instead of going into service, continue with their parents and friends, merely for the purpose of spinning as being a more profitable employment, and [at] which they enjoy more liberty. But there are many who do not like to be so closely confined to spinning and therefore go into service, where only part of their time is spent at the wheel.' In Ireland servant girls seem to have shown enterprise: 'they are generally obliged to spin a dozen of 2 hank yarn in the day besides some drudgery in the house; yet they make out time to spin a little for themselves weekly which in the season amounts to something so as to afford them wearables with other necessary articles. Instead of doing a little for themselves daily, and in order that the wheel might always be fully employed, the usual mode is to give up to five days close application to the employer so to spin six hanks in that time; the 6th day is their own.' (*Statistical Survey, Co. Tyrone*, 1802).

In Wales servants were engaged as much for their skills as spinners and weavers as for their usefulness in the fields. Such a combination of textile work with agriculture was usual in most parts of Europe. In many places the summer was the time when spinning wheels were put away, but in Kilkerry, Ireland, it was said that women preferred to spin than do field work. Another entry in the *Statistical Survey, Co. Tyrone* states 'from the beginning of June to the 1st of September is the dormant part of the year for spinning flax' and suggests that the poor might be employed in spinning wool as 'at that season it is generally procured cheaper than flax. Besides, summer answers better for wool spinning. A spinner can manage flax by firelight only; wool requires more regular light; hence the latter is best subject for the long day. There is another consideration, the root of the common fern is at this season replete with an oily substance which is well known to be an excellent substitute for oil or butter without which wool cannot be manufactured. One pound of wool needs $\frac{1}{4}$lb. of butter to prepare it for spinning. So far as I could learn from the common people the root is cut into

---

*The patron saint of spinners, St Katherine, had no connection with textiles but was sentenced to death by being torn with spokes attached to a wheel; fortunately God intervened and she was beheaded instead. Her Feast Day is November 25th, and in the canton of Graubünden it was forbidden to spin on that day, a prohibition which once may have been more widespread in Europe.

short pieces, bruised in a mortar put into a cloth and pressed out.'

The mistress, daughters and their servants spinning for their own needs would spin their thread according to the cloth required, which led Richard Hall to remark: 'A skilful House-wif's bolt of linen cloth is generally speaking preferable to a master weavers' bolt of the same degree of fineness because she has taken care to give proper twist to both warp and woof.' When there was yarn to spare, even the gentry were not above selling it, since this could be done discreetly through the pedlars who called at the back door. Robert Reid ('Senex'), writing about the late eighteenth century, describes how 'Most families . . . were in the practice of employing customary weavers to execute the weaving of their home-spun yarns; but as these yarns were generally unequal in the grist, itinerant yarn dealers either bought, sold or exchanged yarns with them in order to have the said yarns properly classified and sorted for the loom.' One such dealer in linen yarn was David Dale, who as a young man was 'tramping the country and buying it in pickles from farmers' wives' which led him to be an importer of yarn from the Low Countries (J. O. Mitchell, 1905). Dale later became the father of the cotton trade in Glasgow, where with Arkwright's co-operation he founded a cotton mill at New Lanark in 1783.

McCall tells charmingly of an Earl of Eniskillen (Ireland) who 'took the greatest interest in the production of high-class yarn by his tenants in Fermanagh. He gave premiums to the best spinners; and being himself fond of practical mechanics, he had a wheelwright's workshop erected at Florencecourt, where he frequently spent his leisure hours in repairing the wheels and reels of the peasantry in that neighbourhood. All this recognition of the art by the landocracy not only gave especial dignity to that most interesting department of rural industry, but it called into existence a spirit of emulation which had the healthiest influence on the people. In the parlour of the country squire, by the ingle nook of the snug farmer, and at the fireside of the industrious cottage, hand-spinning was the popular work for the softer sex, and yarns were produced of such fineness and beauty as to become the subjects of universal admiration.'

In the eighteenth century it was quite usual for competitions to be organised and prizes offered for spinning, but if there was an understanding across the social barriers over this common cause, it must not be forgotten that there was a class distinction among the fibres. In the middle of the century at the order of the Brecknockshire Society a scheme was put forward to develop a yarn market at Brecon, and a series of competitions was suggested, mainly aimed at the under 18-year-olds, to encourage the output of the badly needed yarn, in order to get a better return from the abundance of Welsh wool. However, 'There is besides another kind of employ', states the proposal, 'which might be equally beneficial with the manufacturing of our wool, and in some respects, would be even superior, because it might be rendered more universal, and that is the linen manufacture. The spinning of wool is only calculated for the lower sort of people, whereas the other comprehends all ranks and degrees. The delicacy of

71 The Spin House. Illustration from Abraham and St Clara 'Something for All'. After an engraving by Jan Roman de Jonge, Part II

the finest lady cannot be disgusted with the spinning of flax, and as an example is far more prevalent than precept. We have that confidence in the patriotism of ladies, that they will gladly embrace an opportunity of setting a pattern that may be productive of such happy effect.' (The proposals are published in full in *The Welsh Woollen Industry* by Geraint Jenkins.)

A century earlier had seen the introduction of spinning schools. They were set up either as charitable institutions or they were intended to promote industry. The first one we hear of was in 1628 when 24 poor girls were taught to spin, knit and make bone lace at Sir William Borlas's Free School at Great Marlow.* Another seventeenth-century spinning school was started at Lydbury in Shropshire, and there was Thomas Firmin's in London. Andrew Yarraton, in his pamphlet of 1677 suggesting ways of expanding the linen industry in England, gives a lengthy and entertaining description of the strict discipline imposed in a flax spinning school in Germany.

*The thread needs to be very smooth and even for lace; early examples are found to be made of short lengths of linen thread, about 20–22in. (51–56cm.), one would think about the length of the fibre and therefore the draw-out, presumably to avoid having unevenness over joins. In England, Devon and the South Midlands were the two principal districts where lace was made. When Celia Fiennes visited Honiton in Devon she remarked: 'here they make the fine Bonelace in imitation of Antwerp and Flanders lace, and indeed I think it is as fine, it only will not wash so fine which must be the fault of the thread.'

In Scotland the first spinning school was started at Peebles on Tweedside in 1633, the children learning to spin on the small wheel. A year later it was reported that two little girls had learnt to 'Spane at the wisserit quheill' (worsted wheel – Dean, 1930). In the eighteenth century six to eight-week courses to teach flax spinning were run in other parts of Scotland during the winter months so as not to interfere with agricultural work; in many cases the teacher was the wife of the local schoolmaster. In an attempt to introduce flax raising and spinning on the island of Lewis in the mid-eighteenth century a school established at Storna-way ran eight-week courses; the ages of the pupils ranged from 8 to 40 and at the end each received a wheel and reel. Provision was also made there for girls to teach themselves without attending the spinning school, and they too received their wheel and reel plus a bonus of a crown. The little wheel was adapted for flax, according to Mrs Anne Grant, and 'it had many enemies to encounter before it got a footing in the Highlands; which it never obtained till the country was disarmed; and the good women used to speak pathetically of the [17] '46 as the sad era which introduced little wheels and red soldiers into the country'. (In what way it was 'adapted' for flax is not recorded; perhaps simply with the addition of a distaff.)*

Poor relief in many areas often took the form of a spinning wheel, and cloth production was carried out in certain workhouses and houses of correction which gave employment to the poor. London's Bridewell had two distinct functions to perform: on one side it was a house of correction, on the other a technical school for young people – the forerunner of so many others. In Cambridge in 1675 the workhouse, also a house of correction, had a room to employ five wool combers and the poor could find employment by spinning and weaving.

The idea of such institutions was originally continental, German and Dutch, the house of correction being called a spin house (71). In Amsterdam, according to Jan Wagenaar writing in 1760, prior to 1596 vagrant women were sent to the chapel of St George's Church or St Peter's Hospital 'where women were equipped with spinning wheels, reels, cards, and other tools and [they] tried to teach them a suitable trade'; but also in that year the first spin house was built for poor girls to spin wool (Sellin, 1944). However its character changed and when John Evelyn visited it in 1641 he found it just 'a kind of Bridewell where incorrigible and lewd women were kept in discipline'. John Howard making his tour of continental prisons in the second half of the eighteenth century found many houses of correction and workhouses where spinning was carried on. In Bremen they were spinning cow and goat's hair but, he reports, 'a work-house has been

---

*The similarity between linen and a tightly spun worsted yarn (from coarse wool) is pointed out by I. F. Grant. When a hank of yarn made from combed wool was found in the Hebrides in about 1940, even though spinning and weaving still went on there it was not known what the hard yarn was, but it was thought to be linen. Queen Victoria ordered that 'hard tartan' should not be used by Highland Regiments as being too harsh for the men's bare legs and that Saxony wool be used instead. This was the name given to merino wool when it came from that part of the Continent and has no connection with the Saxony spinning wheel.

lately established for the purpose of employing them; and here [in] two rooms, I saw about 170 from six to nine years of age, spinning (with small wheels) under proper masters and mistresses'. In the house of correction in Wurtemburg there was a well-regulated woollen manufactory evidently using great wheels, for Howard reports: 'The wheels were larger than our spinning wheels, the diameter being four feet. The women were all spinning or carding in one large room. As their spinning was of various kinds, there was a room with cupboards where each person's work was laid up separately.' In Vienna the prisoners spun cotton: they saw it weighed out and could take as much as they could spin in a week.

For industry, the production of yarn was organised in much the same way all over Europe, being carried out mainly in the home by women and children and therefore known as the cottage, domestic, or putting-out system. In this way it could be a part-time employment, combined with running a household and tending the land, but for many spinning was the only means of livelihood, rewarded with the most meagre wages and many are the references to the 'poor spinner'. There were, of course, good times and bad times in the cloth industry as the results of wars, political and social struggles, all of which took their toll, and for the spinner at the bottom of the social scale and wage list, this could mean the difference between just enough to eat and starvation. Where there was a cloth industry, there in the surrounding areas, sometimes far afield, would be the industrious spinners. Because they were so scattered it never proved possible for them to organise themselves into spinners' guilds, nor was this encouraged by any other section of the industry, and thus they remained unprotected in this way. The ultimate use of the yarn was of no concern to these spinners, their job being to produce the same type of yarn as evenly as possible and in the least amount of time. The spinners therefore became very specialised in their individual way of spinning, always using the same tool and the same raw material. This they were either supplied with or could procure for themselves.

In the West of England woollen industry the clothier was a businessman who did not do any of the work himself. He would buy the wool, have it washed, dried and oiled and have his servants distribute it to the cottages for carding and spinning. They would collect it the next week and at the same time deliver a fresh lot of wool. Wool was also for sale in the local markets, and there independent spinners could buy the small quantities they could afford and sell their spun yarn. Then there were the yarn badgers or yarn choppers – the chief of market spinners – who traded in wool and yarn through the markets. These middle-men were unpopular with the clothiers, particularly with the small businesses who had to buy in the local markets, since the yarn badgers kept up the price of yarn and some were also cheats. They fumed the wool with hot water to make it weigh more and mixed different sorts of wool in the spinning, causing the cloth to shrink unevenly in the fulling. This would be awkward for the clothier when his cloth was inspected, as the faults would be discovered too late for legal redress against the spinner or the dealer. Early in the seventeenth century the fine, short-stapled wool of the merino sheep or, as it was called, Spanish wool, was imported for use in the West Country industry, and in 1658 a Dutchman was brought over to teach spinning to children in the workhouse at Salisbury (G. D. Ramsay) – which seems strange since woollen spinning on the great wheel would have been an established method in that area.*

In 1686 John Aubrey in his *Natural History of Wiltshire* made a memorandum: 'The art of spinning is so much improved within these last forty years that one pound of Wooll makes twice as much cloath as it did before the Civill warres' (1642). This would surely be because people had mastered the art of using the spinning wheels, as he adds 'In the old time . . . they used to spinne with Rocks; in Staffordshire they use them still'.

In East Anglia, famous for its worsted cloth, the system in the industry was much the same as in the West of England, with the clothier buying the wool and oil and being responsible for getting it to and from the carders and spinners. There were many different types of cloths – wools, worsted, mixtures of the two, silk and wool, linen and wool and so on – made in different widths and each with its particular name. Those known as the Old Draperies were, broadly speaking, thick heavy types of cloth suitable for cold climates, but in the sixteenth century when overseas trade was expanding, types of cloth more suitable for warmer climates were introduced by the Dutch and Flemish refugees and particularly by those who settled in East Anglia. These were known as the New Draperies, in many cases adapted from the Old and lighter in weight, which must have required the spinner to produce a finer thread.

In 1754 a woman of East Dereham produced 'the most extraordinary performance in spinning ever known . . . being twelve dozen and six skeins of curious, hard even spun crape yarn . . . which weighed only sixteen ounces and nearly two drams' (J. James). Another spinner, Miss Ann Ives of Spalding, was awarded a silver medal by the Society of Arts for very fine spinning, so fine that no manufacturer could be found to weave it, except Mr Harvey of Norwich who was constantly endeavouring to find wools sufficiently soft for shawls. In 1788 he wrote to Mr More, then Secretary of the Society of Arts, '. . . the yarn spun so very light as above mentioned from so coarse a material is rather curious than useful; for though it would make stuff very thin and fine, it would not be sufficiently soft and silky. It equals in fineness of thread, the yarn with which the Indian shawls are made: but it bears no comparison to its softness and silkiness.

*In Holland it appears that it may have been quite usual for men to do this type of spinning, or anyway a certain form of it. There are at least two representations of Dutchmen spinning. One is in a seventeenth-century plaque on the Lakenhal in Leiden (illustrated in *Textile History*, 1974); the wheel can hardly be termed great, the top coming little above waist level. The other, a century later, is in the painting reproduced in part on page 31. This is by no means the only time we hear of men spinning, either professionally or as a stop gap in a year which had been bad for agriculture, but more often men's connection with spinning was with the delivery and collection of the materials etc., while the older men and boys helped with the carding and reeling rather than doing the actual spinning.

The wool of Tibet and Cashmere sheep [goat?] from which the shawls are made is soft and silky to a wonder degree; and yet sufficient length to be combed.' The two types of yarn produced by these two spinners, one so over-twisted and the other under-twisted, probably both used the long combed wool of the Lincoln or similar breed of sheep.

Daniel Defoe, always keen to see honest work in progress, wrote while on his tour in 1724: 'When we come into Norfolk, we see a face of diligence spread over the whole country; the vast manufactures carried on (in chief) by the Norwich weavers, employs all the country round in spinning yarn for them; besides many thousand packs of yarn which they received from other countries, even from as far as Yorkshire and Westmorland . . .'.

The organisation of the cloth industry in Yorkshire itself was rather different. The manufacturer was the weaver himself, working in his own house with his family to help; or if in a workshop he was the master weaver, responsible for the business side as well.*

In the middle of the eighteenth century more Yorkshire manufacturers were turning to making worsted cloth. Even a small independent weaver needed more hands than his family could provide to do the spinning, and the organisation of spinners involved a great deal of travelling (already shown by Defoe's remarks concerning the Norwich industry). 'The master weavers either put out the wool themselves to a sort of middle-men, in districts and villages, who again had under them a number of spinners, or else made their purchases from the master-woolcomber, who, often likewise, was an extensive "putter-out" of wool to spin', wrote James. Alternatively shop-keepers acted as agents but they could not be relied on for employing only good spinners since they would not offend a customer by refusing her work. Yarn was often scarce and looms would be idle for the want of the right type of thread. There was much complaint about the lack of uniformity in yarn, particularly as the children's efforts when learning to spin would be reeled into the same hanks as the better-spun thread.

James said that the 'common one-thread wheel' was used up to the end of the eighteenth century for spinning wool and could still be seen in very many farmhouses in the north of England (1857). 'But in the worsted business', he continues, 'there was a peculiarity in yarn spun by this wheel which gave it a great

*The Witney blanket industry was similar to that of Yorkshire in that it was largely independent of intermediaries. It depended mainly on felled wool (wool from dead sheep); Dr Plot in his *Natural History of Oxfordshire*, written at the end of the seventeenth century, notes that the wool came from as far away as Romney Marsh, Canterbury, Colchester, Norwich, Exeter, Leicester, Northampton, Coventry and Huntingdon (all places that grew long stapled wool). He also comments: 'For improvements 'tis certain that the blanketing trade of Witney is advanced to the height that no place comes near it; some I know attribute a great part of the excellency of these blankets to the abstersive nitrous water of the River Windrush wherewith they are scoured (no place yields blanketing so notoriously white as is made in Witney) but others there are again that rather think they owe it to a peculiar way of loose spinning the people have here-about, perhaps they may both concurr to it'.

advantage over mill spun yarn, namely, the thread was spun from the middle portion of the sliver, thus drawing the wool out even and fine'. James quotes an aged manufacturer as saying 'The mother or head of the family plucked the tops into pieces the length of the wool, and gave it to the different branches of the family to spin, who would spin about nine or ten hanks per day.' From this we deduce that quite small lengths of wool were spun at a time and that it was spun from the middle. If a combed staple of wool or a length of tops about 6–8in. (15–20cm.) is folded over the index finger of the left hand and held fairly tightly between the second finger and thumb, the ends of the staple in the palm of the hand, and a few fibres from the fold pulled out to meet the end of spun yarn on the spindle, it is possible to draw the fibres away from the spindle tip while turning the wheel with the right hand, and they will fan from either side of the finger and be parallel as they receive twist. Although not a true worsted thread, a fine smooth thread can be made and there is no build-up of fibres within the hand. It is a technique that works very well on the great wheel and may well be a way it was done in the eighteenth century.

Another domestic branch of the textile industry which needed yarn was hand-knitting. Although there were several centres for it in England, the cloth-producing areas of Norfolk, Yorkshire and Westmorland were amongst the largest. Contrary to the cloth industry, one spinner is said to have produced enough yarn for five knitters (Hartley & Ingleby). Knitting yarn needed to be plyed, and woollen, worsted and jersey yarn were used. The main demand was in the making of stockings, which in the sixteenth century was also an important industry in Jersey and Guernsey. In this connection, an interesting document of circa 1596 found by J. de L. Mann (*Textile History*, 1973) states that 'spinninges of wooll are of three sortes, viz. either upon the greate wheele, which is called woollen yarne, whereof there be divers sortes; or upon the small wheele, which is called Garnesey or Jarsey yarn, bicause that manner of spynning was first practised in the Isle of Garnesey or Jaresey [sic], whereof also there be divers sortes; or upon the rock, which is called worsted yarne by I Edw. VI, because that manner of spynning was first practised in Worsted in the countie of Norf', whereof also there be divers sortes . . .'. It seems likely that the small wheel used in the Channel Islands, with their close proximity to France, would have been a Picardy type spinning wheel turned by hand, since at that period it was a wheel used to spin wool from a distaff.*

Jersey yarn was apparently finer than worsted (as noted in chapter II, p. 61) and spun on a small wheel, yet it is always supposed that the Norfolk spinners continued to spin with the spindle and distaff because the finest thread could

---

*This may also account for that curious remark made by Leake, the cloth inspector, circa 1557: 'About 1528 began the first spinning on the distaff and making of Coxall clothes. . . .' After all, for centuries, as elsewhere, the English had spun with the use of the spindle and distaff. Coxall whites were one of the New Draperies made in Coggeshall, Essex, and in this instance the type of spinning was introduced by an Italian.

be spun in this manner: it seems equally possible that some of them merely refused to adopt new methods. In 1698 Celia Fiennes noted that: 'the ordinary people both in Suffolk and Norfolk knitt much and spin, some with the rock and fusoe (fuseau, a kind of spindle) as the French does, others at their wheels out in the streete and lanes as one passes'. She makes no comment on the sorts of wheel, but it cannot be assumed that, because the spinners were out in the streets, this was the small wheel, since the great wheel with its hoop rim, and probably legless at one end, would not have been too heavy to have been dragged to the cottage door.

The extent to which spinning a worsted or jersey yarn on small wheels was done in England (we have referred to it in Scotland) is not certain. With the introduction of the New Draperies one would suppose that the Flemish and Dutch also brought with them flyer wheels with which to spin the yarn as they appear to have done in their own countries. Dyer, in his poem 'The Fleece' written in the middle of the eighteenth century, describes spinning wheels that are large, small or two-handed (which needs a distaff), as well as the spindle, all for spinning wool. He is careful to point out that when he saw Lewis Paul's spinning machine it was being used to spin cotton 'but it may be made to spin fine carded wool'.

The fibre that was responsible for the Industrial Revolution was cotton. In England the making of fustians came with Flemish refugees and was centred in Lancashire, where there was already (about 1600) an established woollen textile industry. Much of the cotton imported was in yarn form, but small quantities of raw cotton were used for spinning candle wicks; thus in 1692 we find the spinners of cotton amongst those petitioning the London City Council against the setting up of street lighting with oil lamps. Celia Fiennes recorded when she visited Gloucester that, 'here they follow knitting stockings, gloves, waste coates and peticoats and sleeves all of cotton and others spin the cottens'; and T. Rath (1976) found reference to a 'cotton wheel' in a Tewkesbury inventory of 1699.

As well as hand-knitted stockings there was also woven hose, and as early as 1589 the first actual machine for making fabric had appeared in William Lee's frame for knitting stockings. This used a worsted-type yarn and was later adapted to make silk stockings. In 1730 the stocking frame was adapted further for the use of cotton, the fine yarn that was suitable being imported from India. It then needed to be plyed three-, four-, or five-fold to give sufficient strength to withstand the stress of the machinery. It is therefore not surprising to hear that twisting and twining formed a specialised trade in England (Wadsworth and Mann), just as we have already noted (chapter V) that it was in Holland. The centre of the framework knitters was Nottingham, where the spinners of the area were used to handling the long stapled wools and apparently were not able to adapt to the short staple of cotton. However, in the West of England where the spinners were accustomed to the short stapled Spanish wool they were able to

FIGURE 1.—HAND CARDS.
FIGURE 2.—ROVING BY THE HAND WHEEL.
FIGURE 3.—SPINNING BY THE HAND WHEEL.

72 Plate 3 from R. Guest's *A Compendious History of the Cotton Manufacture*, 1823. Two cotton workers (left, spinner; right, rover).

spin the cotton, but since they made a thread thicker than that imported from India it could be used two-fold. Also, being of inferior quality, it enabled the hosiers who worked at Tewkesbury to undersell those of Nottingham and rivalry set up between them in the 1760s.

The method of spinning cotton by hand described by Richard Guest in his *Compendious History of the Cotton Manufacture* (1823) follows an illustration of two women, one making the roving and the other the thread (**72**). His text reads as follows: 'The cotton after being combed or carded between the hand cards, was scraped off them in rolls about twelve inches long and three quarters of an inch in diameter. These rolls called cardings, were drawn out into rovings on the hand wheel. In Fig. 2. the cardings are represented lying across the knee of the rover. From the spindle of Fig. 2. the rovings were taken to Fig. 3. to be spun into weft. In Fig. 3. the roving lies in the lap of the spinner. On the spindle of Fig. 3. the weft was finally prepared for the weaver. In roving, the cardings were drawn out in an angle of forty to forty-five degrees from the point of the spindle; in spinning, the rovings were drawn out nearly in a right-angle. The hand wheel was first used in the woollen manufacture.' It can be seen that a tube was placed on the spindle on which to wind the roving or yarn.

It is, of course, easier to spin a fine thread from a thin ribbon of fibres than from the comparatively bulky rolag, but with this double drawing-out method of spinning cotton, whose short fibres quickly get locked in the twist, how did the rover manage to put in enough twist to make the roving or slubbing cohere sufficiently to be wound on to the spindle without disintegrating, and yet leave drawing capacity for the spinner? If the rover draws-out at an angle of approximately ten to twenty degrees, that is close to the spindle (as indicated in Guest's illustration), only very loose twist is given and a slubbing can be made; the difficulty is to keep its diameter even. If the spinner has her spindle rotating in the opposite direction to that of the rover, then there is no difficulty in drawing-out and a fine thread can be achieved. The twist can be tightened by increasing the angle from the spindle to nearly 90° as Guest suggests. As the spinner winds on the yarn, more roving is paid out from the cop in her lap, and in a manner of speaking this cop is acting as a distaff for the roving. It is one of the reasons, one feels sure, that cotton was usually spun from a sitting position; but also, on account of the lesser tenacity and short fibres, it did not particularly require the same amount of space for drawing-out as wool, so longer working hours could be managed if sitting. When one hears that a rover could serve three spinners and do the carding as well (Wadsworth & Mann) the quickness in making the roving can be appreciated – or perhaps one should say, the slowness of the spinner – for it is cotton particularly, when spun on a spindle wheel, that needs extra twist after drawing-out for strength. Tomlinson, in *Useful Arts and Crafts* of 1858, explains that the slubbing billy, the machine which replaced the work of the rover, 'does not form yarn, for the slubbing has to be twisted in the contrary direction when it is afterwards spun on the mule' (i.e. spinning machine).

Wool was also prepared by machinery in this way, but when spun on a spindle wheel, illustrations of wool spinners from the Luttrell Psalter onwards show not spinning from a cop in the lap but a one-handed technique with the fibres (or rolag) held in the hand. If twice spun it is likely to have been done in much the same way as the Irish spinners do it today (described by Lillias Mitchell), with two rolags joined together and pulled quickly into a roving. The rolag is then dropped and the hand returns to the point of the spindle, the wheel given a sharp turn and the thread attenuated by working up the thread to shoulder level. While this is being done the rolag, left hanging, gently untwists, so the roving remains soft and easily draws-out. Some spinners maintain that by using the momentum of the big wheels, two hands can be used and a fine thread drawn-out between them. There is no doubt this was done, but the method is a slower one. When time was so important and with the long hours worked by the practised spinners, it seems likely that forms of one-handed techniques were more usual.

However, an important concern for the inventors of the early spinning machines was evenly-made rovings, and it is noticeable that Lewis Paul, who took out a patent for a machine in 1738, makes no mention of first drawing a roving on a great wheel; instead he says: 'All those sorts of wool or cotton which it is necessary to card must have each cardfull, batt or roll joyned together, so as to make the mass become a kind of rope or thread of raw wooll. In that sort of wool which it is necessary to combe, commonly called jersey, a strict regard must be had to make the slivers of an equal thickness. . . .' (The joining he refers to was done by placing rolags end to end and rolling them along a board or flat surface with the hand.)

In 1733 John Kay, a Lancashire man, invented the fly shuttle which enabled the weaver to throw his weft from side to side by pulling a cord from a central position, thus speeding up the process of weaving. Although it took some years for this to become generally adopted – in the worsted industry not until the end of the century – it caused the weavers of Lancashire to be even more short of yarn; they would trudge round the countryside to the spinners, often finding other weavers on the same mission, and so presents would be offered to the women to encourage them in their labours. With this increased demand for yarn the need for a multi-spindle spinning machine became so paramount that in 1761 the Society of Arts offered rewards for a machine that would spin six threads with only one person needed to work it.

In 1764 James Hargreaves, himself a handloom weaver, invented the spinning jenny for making yarn from a roving.* His machine was directly inspired by the sight of a spindle wheel lying on its side with the spindle vertical. According to the reconstruction by Aspin and Pilkington from the original patent, the first jennies were made with the wheel rotating in a horizontal plane, though this was soon changed to a vertical position. There were 16 spindles, tilted slightly forward, mounted on a stationary bar at one end of the frame and linked by eight single driving bands, one for each pair, to the wheel turned by the spinner's right hand. The left hand manipulated a movable draw-bar and clamp. Below this, the slubbings or rovings were wound on bobbins mounted in a fixed creel (rack), – though on many later jennies it was made movable. The draw-bar was moved away from the spindles with the clamp open, allowing a length of slubbing from each bobbin to pass through. The clamp was then closed, initially by side cams and finally by the spinner's grip on the clamp. With the slubbings thus 'pinched', the draw-bar was moved further from the spindles, these now being rotated to give twist, while the slubbings were attenuated and formed into threads. As the draw-bar was moved to its furthest point, since this lay beyond the stationary bobbins it meant that, at the same time, new lengths of slubbings were unwound from the bobbins in readiness for

---

*Hargreaves had neither wife nor child named Jennifer or Jane. The nickname 'jenny' has usually been taken to be an abbreviation of 'engine'; in the North of England it has also, however, had general meanings like 'sweetheart', 'drudge', 'donkey' etc. It has also been noticed (p. 34, chapter I) as the name given to the ledge for holding a wool comb.

the repeat process. Extra twist was given to the formed threads for strength and when this was considered sufficient, a faller wire, worked by a lever lifted by the spinner's toe, was lowered on the spun yarns to guide them off the tips of the spindles (i.e. backing-off) to a lower part for winding-on and shaping into cops. This was done as the draw-bar returned, still closed, which simultaneously took forward the next lengths of slubbings for repeating the process.

Richard Arkwright's machine, the spinning frame, appeared a few years later. It took its idea from the bobbin drag flyer wheel and from the making of a worsted-type thread. The four spindles and flyers were mounted vertically with a leather belt below for driving them. The slubbings, on four bobbins placed at the back of the main frame, each passed between four pairs of rollers, each roller of a pair revolving in the opposite direction, so that the slubbing was drawn forward. The lower rollers were fluted, and the upper ones were covered with leather and pressed on to the lower by means of hanging weights. Each pair of rollers moved faster than the last, thereby attenuating the slubbings before they received twist. (This system of roller drafting appeared first in Paul's patent of 1738, already mentioned.) Since the flyers were without an orifice, each thread passed from the front pair of rollers through a guide eye to the rotating flyer giving the twist. Then it passed on to its bobbin, which was checked with a friction band. Thus the spinning frame with its drafting before the twist produced a hard twisted yarn suitable for warp, while the jenny with the draw-out and twist inter-dependent produced a softly twisted yarn suitable for weft.

Not only were these two early machines direct developments of the spinning wheels but it is also evident that the inventors did their best to 'mechanise' the movements of the spinner's hands. Hargreaves used the method of drawing-out the fibres from a roving, and attenuating them away from the tip of the spindle in imitation of the cotton spinner (although these machines were first used for cotton they were quite soon adapted and used for wool). Paul too must have carefully watched the hands of the spinner using the flyer wheel for worsted or even flax, and noted the rolling movement of the thumb as the fibres passed between the two hands – or one hand and the distaff – and over the fingers. This may well have given him the idea of using rollers for the drafting operation. Similarly it may have been in direct imitation of the four fingers that Arkwright was led to use four pairs of rollers initially though he evidently found three pairs sufficient. These early machines can to some extent confirm our understanding of the way spinning by hand had been done previously.

Almost immediately improvements were made to both machines. One of the first made to Arkwright's was to dispense with the hooks on the flyer and to add a wooden heart-shaped cam which moved the bobbins up and down in order to wind the yarn evenly. Thus we have this device on both spinning machines before it became an improvement to the spinning wheel itself.

Inventions of carding machines quickly followed, and from the jenny was developed the slubbing billy, mentioned already. But it was not until the next century that a machine was invented to combine these two processes and so do away with the unpleasant task, done by small children (piecers), of rubbing one length of carding on to the next.

Those that used the jennies in their homes went to the mills for the slubbings and those that continued with the spinning wheel also used them. Vickerman, lecturing in Huddersfield in 1879 said: 'My old grandfather was a carding engineer . . . and he used to tell me about the shepherd farmers bringing to him a few stones of wool to card at the little mill at the foot of the Wessenden valley, and when it was carded they would fetch it home in the carding, rolled up in wrappers, and they used to while away their long winter evenings by spinning on the one-thread wheel, deeming it unsafe to trust both carding and spinning to machinery lest the cloth should not wear well; spinning more than one thread at once on the hand-jenny was called machine work.'

Gradually 'machine work' replaced the spinning wheel in industry, and in some areas this caused great desolation.

> 'Grief, thou has lost an ever ready friend
> Now that the cottage spinning wheel is mute:'

So wrote Wordsworth in 1819 in one of his miscellaneous sonnets, and added a note: 'I could write a treatise of lamentation upon the changes brought about among the cottages of Westmorland by the silence of the spinning wheel . . .'.

William Cobbett was another writer who deplored the effects of the Industrial Revolution on the people of England. In August 1823 he noticed near Wesborough Green in Sussex a woman bleaching her home-spun and hand-woven linen, causing him to write in his *Rural Rides*: 'I have not seen such a thing before, since I left Long Island. There, and indeed, all over the American States, North of Maryland, and especially in the New England States, almost the whole of both linen and woollen, used in the country, and a large part of that used in towns, is made in the farm-houses. There are thousands and thousands of families who never use either, except of their own making. All but the weaving is done by the family. There is a loom in the house, and the weaver goes from house to house. I once saw about three thousand farmers, or rather country people, at a horse-race in Long Island, and my opinion was, that there were not five hundred who were not dressed in home spun coats. As to linen, no farmer's family thinks of buying linen. The Lords of the Loom have taken from the land, in England, this part of its due: and hence one cause of the poverty, misery and pauperism, that are becoming so frightful throughout the country.'

After the War of Independence, the making of textiles in America, now free from English restrictions, was much encouraged and industrialisation got under way. It was in 1828 that John Thorp of Rhode Island patented a ring frame and in the following year another American firm, Addison and Stevens, patented the traveller. The combination of ring and traveller developed rapidly; the drafting is done by rollers and the rotating bobbin is surrounded by the ring. This has a small free-moving piece of wire, the traveller, attached to it. The yarn passes from the rollers through a guide eye positioned above the spindle and is hooked on to the traveller, which is pulled round the ring by the yarn at great speed so that it 'balloons' as the result of centrifugal force between the guide eye and traveller. This gives the yarn twist without having a spindle. The ring moves slowly up and down to spread the yarn evenly on the bobbin, and because of the frictional drag of the traveller as more fibres are delivered from the rollers, it brings about the differential speed between bobbin and traveller, and the yarn is wound on to the bobbin.

The fact that the traveller is pulled round by the yarn recalls the flyer drag so often found on spinning wheels of Central Europe, but it is extremely unlikely that this had any direct bearing on the invention as the spindle wheel and bobbin-drag spinning wheel had certainly had on the inventions of Hargreaves and Arkwright.

Despite all the money and energy that was being put into mechanisation, spinning wheels continued to be used well into the nineteenth century on both sides of the Atlantic; indeed mechanical improvements were devised for the spinning wheel itself to step up its efficiency, but it would seem by and large that most women preferred to stay faithful to what they were used to. Spinning faded away as a fashionable pastime, but remained an important part of day-to-day life, particularly in the more remote areas such as mountains and islands which were less touched by industrialisation and where life followed a traditional pattern. This usually included the making of household requirements, and above all a bride's trousseau, even if the cheap cotton goods had come to replace many things including linen sewing thread. It had not been possible to spin the long fibres of flax on the cotton machinery, and when the preparation and spinning of flax were first mechanised the spinning wheel still produced the finer counts. However, according to Binns, by 1837 in Ireland the factories were destroying the women's business of spinning by hand and they were without employment.

In the Lake District, that very area where the turn of events had been so lamented by Wordsworth, an attempt was made to revive the flax spinning and weaving industry in the 1880s by Albert Fleming, a solicitor from Hertfordshire who inherited a house in Westmorland. The inspiration came from Ruskin, another writer who was deeply concerned by the effects of the Industrial Revolution on country people who had thereby lost self-respect as well as earnings by not having work to do with their hands. Ruskin himself did much to encourage handicrafts particularly amongst the underprivileged.

Fleming had an able housekeeper named Marion Twelves, and in their house was an old spinning wheel which no-one knew how to work. An old villager was eventually found who could remember how to spin, but it is astonishing to learn that no further spinning wheels could be found in the area. A wheel was finally sent from the Isle of Man and a local carpenter was able to copy it, and he made a dozen or so. Under the instruction of Miss Twelves and encouraged by Ruskin, many local women learnt to spin. They were then allowed to take their wheels to their homes and were paid two shillings a pound (or more for fine thread) by Fleming. This was the small beginning of what became a flourishing little cottage industry known as the Ruskin Linen Industry of Keswick – the only known time that Ruskin granted permission for his name to be used.

By the early 1920s the industry was dying out, although one of the last of the spinners, a Mrs Parker, continued to spin until she died in 1969 (she claimed that spinning was a cure for insomnia). F. A. Benjamin's booklet about the industry contains several old photographs of the spinners including one of Miss Twelves herself and another of Mrs Parker, both at work. The spinning wheels were the horizontal type with sloping stock and doubled band drive; the simply-made wheels with felloes had a diameter of approximately 16–18in. (40·5–45·5cm.). What is noticeable is the absence of distaffs: both women spin the flax folded over their left index finger and draw from this with their right hand. Recently an old lady was discovered who had learnt to spin in the Lake District as a child and had been taught this technique. If a distaff were used, apparently it was merely as a reservoir for the flax. This technique of spinning flax over the finger could be linked to the same method as described for spinning long stapled wool on the great wheel (see page 183). It would seem reasonable to infer that in about 1800 (the original old lady was 87 when she was 'discovered' in about 1882) it was a method also found effective with flax: it could even have been a technique transferred from the great wheel to the flyer wheel. The finger, after all, is merely acting as a small distaff and spinning flax this way by-passes the time taken in dressing distaffs.

The fact that the women of Keswick, or at least some of them, adopted this method of spinning flax does not mean that spinning with the use of a distaff had been lost in the Lake District. In a booklet published in 1896, *Notes on Hand-Spinning* by Annie Garnett, who was co-founder of the Windermere Industry, the dressing of the distaff and spinning from it is described. To spin, she says: 'The thread must pass under the second, third, and fourth fingers, and between the first finger and thumb. . . . The right hand, which is generally used for spinning, should be held about three inches from that part of the distaff the fibres are being drawn from, and the left hand should be held under the flax, arranging the fibres in a fan shape, narrowing towards that part which passes through finger and thumb of the right hand.' Annie Garnett tells us that the women of Windermere 'spin flax, wool, and silk, and are able to keep an experienced weaver at work the year round'.

When between the world wars the revival of weaving and spinning by hand led interested people to search for traditional methods still alive in Britain, it was to Scotland and Ireland that they turned. Though flax spinning was no more to be found, wool spinning included both woollen and worsted methods.

The Shetlanders, using the beautifully fine wool that was available from their local sheep, were making gossamer-fine thread for their knitted lace shawls. According to an eye witness, they spun the fibres by using a 'constant rolling motion between finger and thumb' of both hands against the twist to straighten the fibres to be spun.

Although in Ireland the big wheel was still used for woollen spinning, in parts of Scotland (the islands of Harris, and Skye, for instance) the same technique of attenuating the fibres away from the point of twist thereby producing a woollen thread was found as a method used on the flyer wheel, and may be another instance of a technique having been transferred from the great wheel. Elimination of the laborious task of backing-off and winding-on makes it a speedier process and one which can be done sitting. Moreover, in most cases, both hands being free, drawing-out was done with the right hand, which allowed for a maximum arm span, and thus in the same direction in respect to the wheel as on the great wheel. But in Ireland, if a woman had learnt to spin on the big wheel, when she used a flyer wheel she would draw-out with her left hand, this being what came naturally to her. It can therefore be appreciated that there is no right and wrong in the direction of drawing-out the fibres.

At this present time, when there is renewed interest in hand spinning, we are able to pick up the threads of past traditions, some perhaps traceable to the fourteenth century, and learn to use them and adapt them, as they did formerly, to produce yarns for our individual requirements. In this way, if we so wish, we can be independent of the machines and bring back the spinning wheel once more to serve its useful purpose.

Chapter Eight

Practical application

NOTES FOR THE BEGINNER ON USING THE FLYER SPINNING WHEEL

A comfortable chair at the right height for YOU is essential.

Wear an apron, particularly for fibre preparation.

First learn to treadle; place the whole of the right foot on the treadle. Push the wheel in the direction you mean it to go with the right hand between the spokes (never start the wheel by moving it backwards a little first); pick up the rhythm with the foot and do not let the ball of the foot and the toes leave the treadle. It should feel as though the wheel is pushing your foot, not vice versa. Practise treadling at various speeds in both directions, stopping and starting; it is possible to stop the wheel so that the crank is just over its highest point in the direction it is to be turned, making it unnecessary to re-start it with the hand. Until this is mastered, place the wheel in position by hand before attempting to spin (73).

73 Crank in correct position to commence
treadling for Z twist

The position of the tensioner should be adjusted so that there is a feeling of draw-in, but not too fast. Have several feet (2 or 3 metres) of starter thread tied to the bobbin and threaded through the orifice. Hold the end in the right hand and adjust tensioner with left hand: treadle, and if no draw-in, tighten; if too much, ease the band. (Remember that the starter thread must be twisted in the same direction as the intended spinning.)

Keep the wheel axle and spindle well lubricated with a non-sticky oil such as machine or motor oil. Leather bearings may also be oiled but not nylon ones.

Spinning needs perseverance in the early stages. Each technique has its own rhythm and each hand has its particular job to do. Prepare a quantity of fibres before starting to spin and keep them to hand, otherwise the rhythm of the spinning is constantly interrupted. There is no need to stop the wheel to join in a new lot of fibres.

MAKING A WOOLLEN YARN

Wool

Choose a wool which is 3–5in. (7·6–13cm.) staple length (not longer than the width of your carders). Only spin wool 'in the grease' if it is in good condition and feels greasy; if sticky, dry or dirty (or if you wish to dye a fleece) scour before spinning.

Scouring

a   Soak wool in hand-hot water for about half an hour, but do not agitate it.

b   Gently squeeze out water and remove wool.

c   Change water (never run water from the taps on to the wool) keeping the temperature the same; add water softener if necessary and a good measure of soapflakes or a bleach-free washing agent. Enter wool and soak for approximately 20 minutes. Repeat stages b and c several times until the wool looks clean.

d   Gently squeeze and rinse in three changes of water keeping the temperature even.

e   If a spin-dryer is available, place wool in a stockinette bag or old pillow case and spin-dry for ten seconds.

f   Dry away from sunlight. Quick drying artificially is quite satisfactory.

Oiling

Use corn or olive oil (test on a sample that it will wash out leaving no smell); this can be rubbed on the fingers and rubbed into the wool while teasing.

Alternatively use a water-soluble spinning oil or emulsion; it can be sprayed (use a scent bottle or insect spray) on to the wool or on to the carders. Some spinners prefer to oil a few days in advance of spinning so that the wool gets well impregnated with oil, in which case roll up in an old sheet and store in a warm place.

(An emulsion can be made with equal parts of olive oil and water, adding a very small quantity of a soda and warm water solution. However, be sure to wash the hanks of yarn soon after spinning, as the soda, in time, can damage the wool.)

Teasing

Take a handful of wool, open it up by pulling apart with the fingers; remove any lumps, vegetable matter, second clip ends (see Appendix, page 229) and then thoroughly mix the fibres.

Carding

Label carders left and right and always use them in the same hands as they will then wear better.

N.B. Left-handers should reverse these instructions since the hand on top does most of the work and it is more comfortable to use the stronger one.

a   Place left hand carder on knee and fill with small tufts of wool from a handful, arranged evenly across the carder; allow no ends over handle end, but they can hang off at the other end, the tip (**74**).

b   Hold carders as shown in plate 75 (the index finger of the right hand may be placed on back of carder if preferred: see Luttrell Psalter, plate 16). It is important that the lower carder rests firmly on the knee. There is no need to change the grip till stage **i**, although there are other equally good methods of using carders when you do so.

74 Carding (a), placing wool on left hand carder

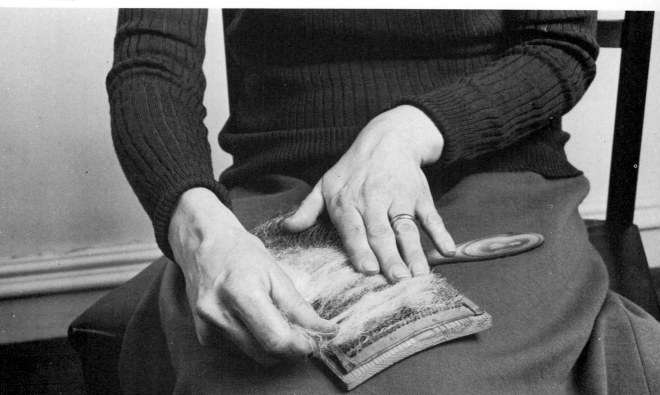

196

75 Carding (c), commencing

c Using only the tips of the carders to start with, draw the right hand carder gently over the fibres (**75**). With each stroke progress further across the carders until the full width of both is being used and all the wool is stroked by the teeth. Use the right hand lightly, particularly if it is fine wool, and do not 'dig' into the fibres. Make sure the ends of the fibres are clear of the teeth after each stroke. Continue only until the wool looks fluffy and airy. Some, but not all, will transfer on to the right hand carder.

d Turn right hand carder so that it faces upwards.

e Carefully laying any fibre ends from the tip of the left hand carder on to the teeth at the handle end of the right hand carder so that there is no possibility of them becoming curled over, sweep the left hand carder down the right hand carder with a quick movement (**76**), transferring all the wool on to the right hand carder. It should then look like a flat sheet of wool.

f Repeat stage **c**; the fibre ends will be at the tip of the right hand carder but some wool will transfer itself on to the left hand carder.

g Repeat stage **d**.

76 Carding (e), sweeping left hand carder down right hand carder

77 Carding (h), laying the fibres carefully on to the left hand carder before sweeping it up the right hand carder

h Carefully laying any fibre ends from the tip of the right hand carder on to the teeth at the handle end of the left hand carder (**77**), sweep the left hand carder up towards the handle of the right hand carder. All the wool will now be on the left hand carder. Repeat stages **c** to **g** once, then stages **c** and **d**.

i To remove the sheet of carded wool from the teeth, sweep the left hand carder down the right hand carder, but as it nears the tip draw it back again, but so that the teeth of the two carders are not engaged and the wool is eased off, with both carders facing upwards (**78**). Often it needs to be transferred several times from carder to carder, keeping these facing upwards, before the sheet of wool is fully eased off the teeth.

j Turn over the left hand carder and transfer wool on to the wooden back; press fibres down with the back tip of right hand carder and either form a roll by rolling between the backs of the carders, or roll with the fingers, in either case towards you, keeping the fibres at the ends well in to prevent the rolag from becoming elongated. Give a final roll between the carders (**79**). The rolag should be of equal density throughout; hold up to the light to test.

78 Carding (i), easing-off the sheet of wool

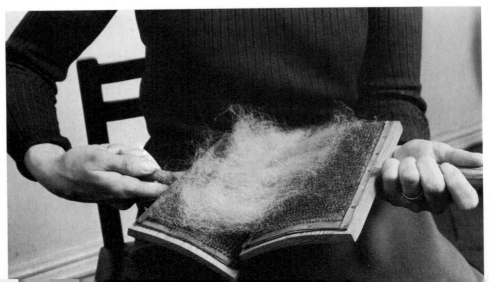

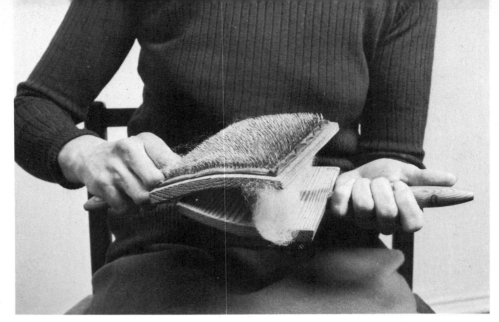

79 Carding (j), the final roll

Alternatively two rolags can be made from one carding; some wools readily divide more than others and it is advisable to fill the carder a little fuller in the first instance. It is easier to remove the first layer while the left hand carder is swept up the right hand carder rather than down it. Make the rolag with the second layer still embedded in the teeth of the left hand carder; then ease off the second layer and form into a rolag.

The amount of carding is, of course, at the discretion of the spinner, but to over-card is both a waste of time and energy as well as harmful to the fibres. The aim should be to produce a firm airy cylinder to make a warm air-trapped yarn.*

Spinning

In the following description of spinning a Z twist yarn the fibres are controlled by the right hand and the spin, or amount of twist, by the left hand. (This long draw method is sometimes found a little awkward for owners of vertical spinning wheels, particularly if the flyers are rather high off the ground. The spinner may find it more comfortable to sit with the right foot on the treadle and then swing the body away from the wheel so that the orifice is pointing towards the left hand, which would then control the fibres while the right hand controls the twist. Conversely, treadle with the left foot and swing the body to the right. Then proceed with hands as follows.)

a Joining: make sure the crank is in the correct position. To join the first rolag to the starter thread, pull some fibres from one end of the rolag and hold this in the right hand with the fibres close up against the thread which is held in the left hand. Start treadling (keeping it slow till stage **g**) and the fibres will catch on to the thread (**80**). Immediately move right hand close to left hand and as fibres form a thread release the end of spun yarn so that they become one. Hold

*To clean carders, take one in each hand with the handles pointing downwards and the teeth facing, and brush downwards, first one and then the other, with a series of short jabbing strokes. When the backs get too greasy for comfort, scrub with soap and water without wetting the teeth.

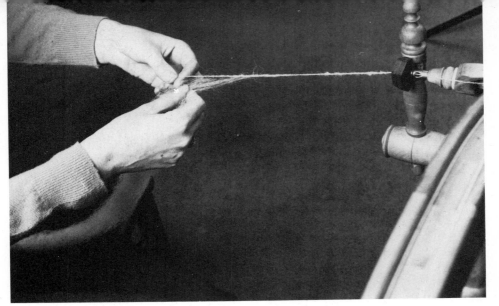

80 Woollen spinning (a), joining

the join with the left hand and draw the right hand a little away from it, at the same time allowing some fibres to slip through the fingers which then receive twist when the left hand releases its grip on the join. Let the join advance towards the orifice.

Alternatively: Divide about the last 3in. (7·6cm.) of the starter thread into two sections. Place some fibres, pulled out from one end of the rolag, between the two sections and hold them all together between the index finger and thumb of the left hand. Start treadling and when a little twist has built up draw right hand away attenuating a small portion of the rolag and release the twist over the join.

b   To start the long draw, grip the yarn between fingers and thumb of left hand (it is immaterial which face upwards) about 9–10in. (23–25·5cm.) from orifice (and to the right of the join) where it meets the rolag, the last point of twist. The left hand does not move from this position.

c   Grip the rolag with right hand about 2in. (5cm.) along; treadle and build up some twist between orifice and left hand without the hands moving (too much twist will cause the thread to snap, but too little will not support the fibres as the draw-out is started). Draw a little away from left hand (**81**).

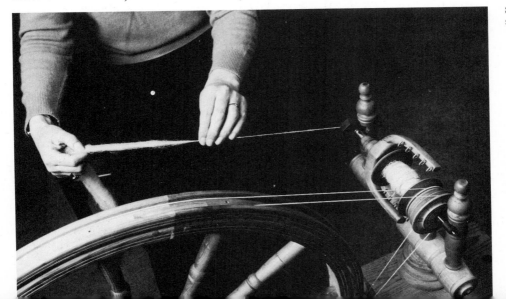

81 Woollen spinning (start of the long draw

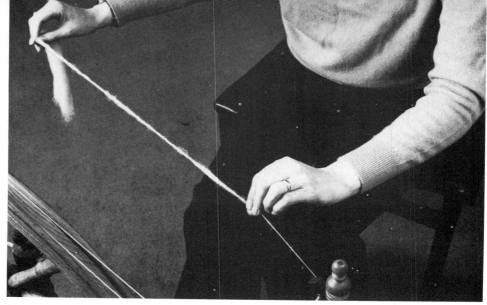

82 Woollen spinning (d), fibres partially attenuated

d   Release the twist into the fibres by slightly opening the thumb and index finger of the left hand at the same time draw the right hand away so that the fibres are attenuated as the twist enters them (**82**). (To the beginner it can seem that a surprising amount of 'pull' is needed.)

e   Continue to draw-out the fibres to the full extent of your arm (**83**). If there is a danger of the thread disintegrating, which might occur if there is not suffi-cient twist to support it, close left hand fingers and thumb on to the thread to build up a little more twist, then release it into the thread.

f   When yarn is formed, give it a little more twist for strength but keep a tight grip with right hand to prevent any twist reaching the rolag.

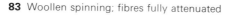

83 Woollen spinning; fibres fully attenuated

g   All this time the draw-in has been resisted, first by the left hand and then by the right hand. Now let the yarn be drawn-in as fast as possible on to the bobbin, increasing the speed of treadle, without the left hand touching the yarn, so that the fibre ends are not smoothed into the twist and the yarn is hairy, and the maximum natural springiness of the wool is retained. The right hand stops about 9–10in. (23–25·5cm.) from the orifice and the hands change places to repeat the process.

It takes practice to gauge the correct amount of draw-out to the amount of twist. The left hand acts like a pincher, opening to let more twist through, or closing if the twist starts to lock the fibres before they have been sufficiently drawn-out.

Difficulties

a   Over-twisting: caused by lack of confidence in making the right hand work fast enough with the draw-out. When this happens, the corkscrews that form catch on the orifice or hooks and obstruct the draw-in. This causes much frustration to the beginner who forgets to look at the formed thread. Moving the yarn regularly from hook to hook on the flyer so that the bobbin is filled evenly, and slightly tightening the tension after each layer is completed, must not be forgotten (see chapter III). A bobbin over-filled in one place causes loss of tension on the yarn and will lessen the draw-in, again resulting in over-twisting.

b   Under-twisting: usually caused by a lazy left hand not pinching the yarn again to build up more twist when there is a danger of the thread disintegrating. Do not attempt to join two half spun threads. Either break off, or start from the other end of the rolag.

c   Driving band jumping off: caused by jerky treadling or stopping the wheel suddenly with the hand.

d   Co-ordination between hands and foot: practise working stages **e** and **f** without treadling, but do not let this become a habit.

e   A slub of unspun wool forming just behind the left hand: due either to too little build-up of twist before the first draw-out is made, or to an insufficiently quick draw-out. If noticed in time, this slub can be removed by rolling it between the left hand fingers and thumb against the twist, at the same time giving a little pull with the right hand.

f   Not achieving the long draw: this is not always realised by beginners. Instead of attenuating the 2in. (5cm.) of rolag, the right hand moves away from the left hand while all the time allowing small amounts of fibres to slip through the fingers from the rolag. This is not the long draw.

MAKING A WORSTED YARN

Wool
Many types and lengths of wool can be spun in this manner, but for the beginner choose from one of the lustre wools using a staple length of 6–8in. (15–20cm.).

Teasing
Take a staple or lock at a time and tease it gently open, particularly the tips, keeping the fibres in their natural alignment.

Combing
Ideally the wool should be combed between two heavy combs, in which case only the minimum of teasing is necessary, but few hand-spinners working in their own homes have the equipment. For those who have, refer to the description on p. 34 (chapter I).

Alternatively, use a dog comb, preferably one with two rows of metal teeth,

84 Combing (a), flicking a staple of wool using a dog comb

clamped to a table. (The plastic variety with two or three rows of teeth is, however, satisfactory.)

a Working a teased staple at a time, hold it firmly in one hand. The wool is gripped, not only between thumb and index finger, but between third and fourth fingers also. This is not always possible on shorter stapled wool. Flick over the comb and draw the fibres as gently as possible through the teeth (**84**); it is usually necessary to flick several times to remove knots and short fibres. Do this from both ends of the staple; the fibres will be aligned parallel and the short fibres remain in the comb. Remove these noils between each combing to avoid clogging it.*

b Make a pile of combed staples with the tips all facing in the same direction.

Alternatively: a staple may be split in half and laid end to end to mix the direction of the fibres.

c Draw the staples into a continuous roving, letting the fibres slide past each other without getting out of alignment.

Alternatively: roll the staples into a bundle (they can be loosely tied, if liked) keeping the fibres parallel, *or* spin from the individual staple held in the hand.

Spinning
In this description of making a Z twist yarn, the left hand controls the twist and the right hand holds and controls the fibres; it may be achieved equally well with the hands working the other way round.

Treadling should be kept to a steady rhythm all through the process, but the beginner should try to keep it slow.

If spinning from an individual staple or a bundle where the direction of the fibres has not been mixed, it will be found that the fibres adhere to one another more readily if the clipped ends face the orifice. In this direction the scales on the wool are flattened into the thread along with the fibre ends, which all helps to make a smoother yarn. In keeping the fibres parallel, they slide alongside each other, so it is usual to give a fair amount of twist to worsted yarn.

a Joining: lay the spun end of yarn from the orifice alongside the staple of wool or roving and include very few fibres in the first draft.

b The left hand works about 6in. (15cm.) from the orifice with the fibres passing over the fingers – they are held firmly by the thumb a little way along the index finger. The right hand holds the bundle of fibres or the beginning of the roving with the fibre ends pointing towards the orifice.

---

*In Australia and New Zealand a flick comb achieves the same result. A miniature carder approximately 3 × 5in. (7·6 × 13cm.) with coarse wide-apart teeth is held in one hand and flicked across first one end and then the other of the wool staple which is held in the other hand. The comb is worked away from the body.

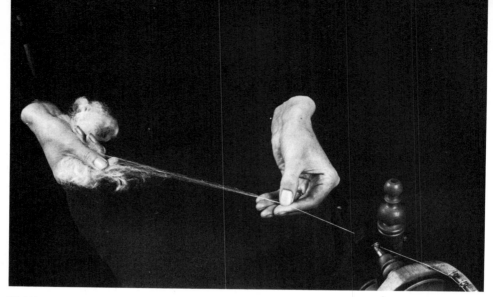

85 Worsted spinning (c), drafting

c   As treadling commences, the right hand with the fingers facing upwards draws back from the left hand allowing a few fibres to slip through the fingers; because the twist is being firmly held back by the thumb of the left hand pressing on the index finger, the fibres are drafted without twist and therefore this can only be to the extent of their length (**85**). The fibres form a narrow fan from the right hand to the left hand.

d   As the thumb of the left hand releases its hold sufficiently for the twist to enter the fibres, the right hand moves forward towards it; at the same time the wrist turns the fibres into the twist while the left hand thumb rolls gently over the thread as it forms, to smooth down the fibre ends into the twist (**86**).

e   As movement c is repeated the left hand thumb rolls back to its original position catching in a few fibres while the right hand moves away and turns back to its original position. At no time does the left hand lose contact with the fibres and thread. Inevitably, after each staple is spun, a small quantity of short fibres remain in the hand and these should be discarded.

86 Worsted spinning (d), smoothing fibre-ends into the twist

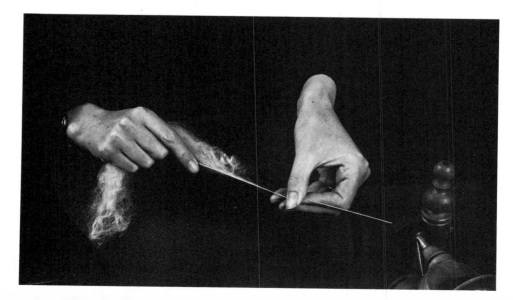

For an S twist, the direction of the rolling movement of the left hand thumb and the turning of the right hand wrist must be reversed.*

Difficulties

a   Mastering the rolling movement and gauging the distance of the draft without twist. This comes with practice.

b   A build-up of wool in the right hand: this happens more often if the fibres are not very long. They must lie along the palm of the hand and not be gripped so hard that they cannot slide, when they get jumbled up.

c   Making the fibres adhere: either insufficient twist, or the tips facing the orifice instead of the clipped ends.

d   Fluffy joins: if the front hand leaves the yarn and fails to smooth over the join. This is not a problem if spinning from a roving.

THE SHORT DRAW

This can be used in many ways according to the result required.

Preparation

Wool can be carded but rolled across the carder so that the fibres lie more or less parallel (87).

87 Rolling carded wool across the carder

*It is not usual in Britain to spin wool from a distaff but in Scandinavia this is done. It is particularly useful for the Norwegian *Spelsau* wool which, like some of our mountain breeds, has an undercoat of short soft wool and an outer coat of hair; but unlike the Scottish Blackface, for instance, the hair is lustrous. It is therefore useful to separate the two, by pulling gently between the two hands from both ends of the staple, the lustrous hair being suitable for tapestry weaving. The fibres are aligned parallel and rolled on to a spindle-shaped distaff (about 12in., 30·5cm.), the fibres lying vertically round it and drafted from the bottom, in the worsted manner.

Alternatively, use a drum carder which is similar to a miniature carding machine and is turned by a handle (**88**). It is a great time-saver and is therefore a useful piece of equipment particularly for hand-spinners needing carded wool for making thick rug wools, mixing the wool and hair of mountain breeds, and also for very short fibres which are difficult to card.

Fibres should be clean and well teased; place on the tray and feed evenly to the small roller (the licker-in) with left hand while the handle is turned with the right hand. The fibres become transferred on to the large drum and after several turns of the handle are ready to be doffed. A strong piece of wire (or steel knitting needle) is slipped under the lap of carded wool at the point where there is a join in the carding cloth and gently lifted up to ease a break in the lap. Peel the lap off by holding one end of it with the right hand while slowly reversing the drum by the handle with the left hand. The distance between the two drums (rotating in opposite directions from each other and at different speeds) can be adjusted according to the type of fibre being used; it is easier when doffing to adjust them further apart. As the first roller does not distribute the fibres evenly on to the drum, it is more satisfactory to place very short fibres directly on to the drum.

The fibres are arranged roughly parallel along the laps, which can be split up and either rolled into rolags with the fibres lying lengthwise or crosswise, or made into a continuous roving.

The laps are very convenient for oiling, first on one side and then the other. It is also a good arrangement for placing in layers in a dye-pot.

Alternatively, long wool can be teased as for worsted spinning – but not carded or combed – and spun in the grease. If the tips are sticky the staples can

88 Drum carder with tea
brown wool on the tray

89 Short draw

be steeped in cold water for about an hour and dried. A little oil rubbed on to the fingers can facilitate spinning.

Spinning

a Whichever hand holds, and therefore controls the fibres, it should move away from the hand controlling the twist (**89**).

b If the fibres are arranged parallel and the rolling movements are used as described in the last section, a smooth thread can be made with most lengths of fibre. However it does not necessarily need so much twist, since the short fibres have not been removed, so cohesion is easier. This may be termed a semi-worsted yarn.

Alternatively, the front hand can simply smooth the fibres as the thread forms, *or*, if the front hand controlling the twist releases the thread after each draw-out and allows it to be wound on to the bobbin without touching it, a much more woolly thread will be achieved. If the hands work fast in relation to the treadle, there will be less twist and a softer yarn will be made.

With wool the hands can work quite far apart, but with some of the more slippery fibres it is necessary for them to be closer together.

Difficulties

The most usual is over-twisting and the making of a rather hard yarn. To overcome this, the hands need to work faster in relation to the treadling speed.

Uses

a   Spinning straight from the fleece: a portion of good condition fleece is placed on the floor beside the spinner or held in the lap. It requires a certain amount of teasing with the fingers as one goes along and the right hand has to work hard to produce the right proportion of fibres. Spinning at random from coloured fleeces (i.e. browns, greys etc.) can make very effective yarns.

b   Some fleeces suitable for woollen spinning have special colour effects (such as brown with pale tips) which would be lost if carded.

c   Angora rabbit: the hair is plucked and a good quality is about 3in. (7·6cm.) in length. The fibres are slippery but no preparation is necessary, and if spun from a handful with no particular arrangement of the fibres, and the left hand does no smoothing movement, the yarn gets a fluffy light appearance. It makes a useful knitting yarn either plyed or mixed with a wool of similar staple length.

d   Dog hair: it is best to use the combings. Each breed has its own characteristics. Samoyeds produce a soft white hair while the Old English Sheep Dog has a more woolly hair, as have poodles. The Pyrenean Mountain Dog, Chow, Pekinese, Sheltie, and Spaniel all produce hair suitable for spinning, some more slippery to handle than others, their variety of colour being one of their interesting attributes. The more slippery types can be mixed with wool, which makes them more manageable and easier to spin. Dog hair can be scoured, oiled and carded or, if clean and in good condition, spun from a handful.

e   Cat hair is not successful on its own since it matts so much, but if required for its colour should be mixed with wool.

f   Mohair, the hair from the Angora goat, is a strong silky lustrous fibre and very white. It can be scoured, and because of its silky texture can be spun without re-oiling. If spun in a worsted manner it makes a heavy strong lustrous yarn, but if carded yet not rolled, and the fibres drafted from mid-way along one side, a fluffier yarn can be achieved and can keep a lustrous core. Fast movements with the hands are necessary to avoid over-twisting and it is one of the few occasions where the best effects can be achieved by pulling the fibres towards the orifice with the left hand.

g   Machine-combed tops are sometimes the only way in which some of the more exotic fibres, such as camel (the inner coat, the outer being very coarse and spikey), alpaca, cashmere, also mohair, can be obtained. Man-made fibres come in tops and also some of the coarser hair such as is found on the domestic goat. To spin from tops, break off (by gently pulling) about 24in. (61cm.) of the roving; peel it lengthwise down the fibres several times until the pieces are reduced to a size only a little thicker than the required finished thread. The fibres are spun parallel with a short draw and will give a smooth yarn with a tendency to be hard unless the hands move very fast. This is probably the best way to deal with coarse fibres and may be satisfactory for certain uses with any of them, but the quality of a soft fibre can easily be lost.

Alternatively: to retain the softness, break off about 4in. (10cm.) of roving, open it up sideways so that it forms nearly a square of fibres and spin from the side, as mentioned above for mohair.

Alternatively: open up 4in. (10cm.) lengths of roving, roll into rolags at right angles to the fibres and spin using the long draw method.*

COTTON DRAW

Many hand-spinners do not have much opportunity for choosing a grade or staple length of cotton, but a long one (i.e. 1–2in., 2·5–5cm.) and fine fibres are easier to handle and give the best results.

Preparation

Ginned cotton is compressed into bales so needs to be opened up, freed from dirt and fluffed up before spinning.

Alternative methods:

a Bowing: a bow can be made with a flexible stick, approximately 30in. (76 cm.) long, bent into a curve by a length of gut (such as a violin A or D string) tied to both ends. (If longer lengths of gut are available larger bows can advantageously be made.) Pluck the gut vigorously over the pile of cotton; some will adhere to the gut and must be eased off, but with perseverance the vibrations will fluff up the cotton. It can then be divided up and rolled into rolags or carded.

b Steaming: a few minutes in a colander over boiling water, but care must be taken not to wet the cotton.

c Beating: use a whippy stick and move it vigorously in a pile of cotton placed on the floor.

d Carding: the carders should have finer teeth than those used for wool; only the tips of the carders need be used, and very few strokes.

Spinning

a Hold the rolag in the right hand and pinch firmly between thumb and index finger. As treadling commences allow a small fan (not longer than the length of the fibre, about $\frac{1}{2}$–1in., 1·3–2·5cm.) to slip through the fingers and immediately form a thread.

b The left hand is positioned a few inches from the orifice and opens and shuts on to the thread to give twist or control it, as necessary.

c The hands can either work close together, or the right hand can move away from the left hand, but still keeping a firm control on the little fan and the continually forming yarn (**90**).

---

*Machine-carded wool, known as slivers, can be broken off into 24in. (61cm.) lengths, peeled down, to make rovings. These can also be spun using the long draw method attenuating portions of the roving between the hands.

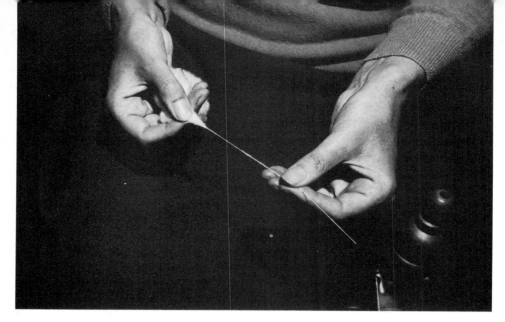

90 Cotton draw

d It is sometimes advisable to ease the driving band so that the thread is not pulled on to the bobbin too quickly.

This is a useful method for any very short fibres, though it will not produce such an elastic or warm thread as when woollen spun.

Difficulties

a Under-twisting: (i) For the spinner used to spinning wool, difficulty is sometimes found when handling such very short fibres and the thread is constantly pulled out of the hand and disintegrates. Slacken the driving band, and every now and then let extra twist run into the yarn. (ii) To begin with it can be difficult to judge when the thread is strong. It may look coherent, but when reeled off, disintegrates. It of course needed more twist.

b Over-twisting: the same advice applies as given for the short draw; or the driving band is too slack.

c Producing a fine thread: A very firm control is needed between index finger and thumb to let only the smallest fan through the fingers. The easiest way to make a really fine thread is by using machine-prepared rovings; this underlines the importance of well-prepared fibres for such types of spinning.

A NOTE ON SILK

Silk is available to hand-spinners mainly in the form of tops. Pull off small lengths, about 6–8in. (15–20cm.), and peel these into several narrow strips. Fold over each strip and spin from the centre of the loop using either the cotton draw method for a fine thread, or the long draw for a bulkier one.

Strusa:

This form of silk waste arrives looking like cream-coloured straw. First de-gum; add the bundle of strusa to a generous portion of soapflakes dissolved in warm water. Heat to just below boiling point and keep at that temperature (195°F or 90°C) for 1½–2 hours. Rinse and repeat the process once or twice. Finally rinse and lay out to dry. A great deal of patience is required to sort out the tangles in the fluffy mass of fibres. Have several cardboard spools to hand and when possible wind off lengths. From these, cut off 4in. (10cm.) lengths and card very lightly. The really fluffy short fibres can be spun by the cotton draw method, but the longer ones can be spun in the short draw method. Although the yarn lacks the lustre of the machine-prepared tops, softly spun it can make a delightful weft yarn. Also, it often forms into little nops as it is spun, which makes an interesting yarn.

SPINNING FLAX

Flax fibres can vary from coarse and brittle to fine and supple. Flax line can be purchased in stricks with the two ends of the bundles tied loosely together. It is ready hackled but the fibre lengths are not always uniform.

Preparation
Untie and identify the root end where the fibre ends should be evenly together and should look woollier, and feel coarser than the tips, which are finer and uneven. Hold the strick in one hand by the root end and separate off a finger, 1–2oz. (28·4–58g.), with the other hand, pulling away rather sharply. (Larger quantities at a time can be used, but the beginner should take small amounts to gain practice in dressing the distaff.) Still holding it from the root end, give the **2oz. (58g.) finger a mild shake; only if it sheds a lot of shorter ends is it necessary to hackle.**

Hackling
In the lack of a fine hackle, use a dog comb clamped to a table. Pull the fibres through the comb, first from one end and then from the other. Because the fibres are so long, wind the half which is not being hackled round the wrist of the hand doing the pulling, while the other hand is placed on top of the fibres close to the comb, lightly pressing to guide them through the teeth.

Dressing the distaff
There are many methods of doing this, and the type of distaff can influence the method chosen, but as the flax line available today is usually quite long, some sort of crossed arrangement of the fibres makes them easier to handle. How closely they are attached to the distaff is a personal matter, but to spin a fine thread the flax is more easily controlled if wound on firmly and well supported.

Figure 14

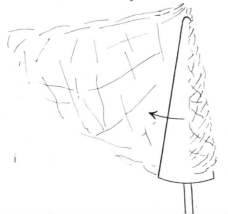
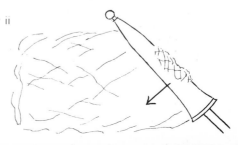

Rolling flax on to a distaff:
i. a semi-circle of prepared fibres on to a cone-shaped distaff.
ii. an oblong of flax diagonally on to a truncheon-shaped distaff.

91 Dressing the distaff (method I, b); spreading out a fine film of fibres and folding over to start the next layer

Method I

a   Wearing a large apron, sit in a chair, the distaff within reach. Tie the root end of the 2oz. (58g.) finger of flax with a half knot in the centre of a piece of string long enough to tie round your waist with a bow at the back.

b   Take hold of the flax in the right hand and swing it (pivoted from its tie at the waist) as far round as possible to the left. With the left hand, pull out a fine film of fibres, both hands moving slowly across the lap to make a semi-circle of fibres. When the right hip is reached, change the flax into the left hand and fold it over to return (**91**), then doing the same thing from right to left, the right hand now pulling out the film of fibres and guiding them with the flat palm of the hand which also smooths and flattens them. The bottom of the fibres will sweep round even if they are out of reach to be guided, but care must be taken that they are being properly fanned out near the waist tie.

c   When all the flax is used up, untie the string, both from yourself and from the flax (or cut it, leaving the knot round the flax if you prefer), and roll the semi-circle of fibres on to the distaff (the stick pointing away from you), starting at one edge of the semi-circle and smoothing the fibres round the distaff as you go (*figure* 14 i).

iii. an oblong of flax across on to a pole-type distaff with small lozenge shaped cage at the top
iv. a U of flax on to a spindle-shaped distaff

iii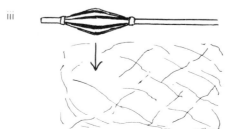

IV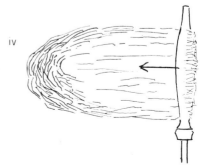

d   Have 2 yds. (183cm.) of 1in. (2·5cm.) wide ribbon, made of a smooth material that does not catch on the flax. Tie at the middle round the top of the distaff with a half knot and spiral round and down the dressed flax. Tie at the bottom with a bow.

(This is a well-known method of dressing the distaff and is described by Horner, p. 99, chapter IV, as used in Ireland and coming from Holland.)

Alternatively: Kneel on the floor spreading out the apron and make the semi-circle round you: *or* with the apron on a table make the semi-circle.

Method II

a   Working at a smooth table top (it is easiest to stand) divide the 2oz. (58g.) finger into approximately eight small lots.

b   Take each one separately and spread it into a fine film making an oblong layer of criss-crossed fibres. Place aside.

c   Continue to make layers in this manner, piling them on top of each other the oblongs all lying in the same direction.

d   Roll the distaff diagonally across the oblong (*figure* 14 ii) and tie with ribbon. (A method described by Annie Garnett in 1896.)

Alternatively: roll the distaff straight across the oblong. This method can be used for the pole-like distaffs (*figure* 14 iii), although the flax can be placed on this type by just shaking out and arranging carefully round the pole, then tying firmly with ribbon.

Method III

Take the 2oz. (58g.) finger of flax and loop it round to form a U. Gently pull the fibres out from the centre of the U, keeping the shape, at the same time rolling the fine film of fibres on to the distaff. This is easiest done on a long table and is particularly useful for small spindle-shaped distaffs (*figure* 14 iv) (often found on spinning wheels from Scandinavia, where this method is used). It is not necessary to tie on with ribbon, but the fibres may be smoothed with a dampened hand.

Preparing to spin

Whichever method is used, when the distaff is in position, whether free-standing or attached to the spinning wheel, the bottom of the fibres should be at a slightly higher level than the orifice. Place the distaff either on the left of it or above it. A little pot of water or a dampened sponge is best placed to hand for wetting the fingers, avoiding the need to interrupt the rhythm of spinning. It is usual, though not essential, to spin flax in an S direction for reasons already mentioned (p. 18, chapter I).

Spinning

The similarity in the hand movements between making a fine smooth linen thread and a worsted one has already been pointed out (p. 89, chapter III). Flax is usually given more twist than wool, so the smoothing motion can be quite leisurely. Either have a slightly slacker driving band or a spinning wheel on

92 Spinning flax (method I,b)

which the circumferences of the two whorls are quite close.

For a thicker yarn less twist is given to avoid the result being like string. A rougher thread can be made from spinning flax dry.

Method I

a   Joining: have a longish length of S-spun starter yarn from the orifice so that it can be placed round the dressed flax. As treadling commences, draft in a very few fibres from the bottom of the dressed flax with the starter thread, using the dampened thumb and index finger of the left hand, with the thumb placed at the extreme tip of the index finger.

b   Between the point where the formed thread passes round the third finger and the orifice a tension is kept, thus holding back the twist while the fibres are being drafted (**92**). The hand does not get closer to the orifice than about 9–10in. (23–25·5cm.). As a few fibres are selected and drawn down, the wrist is slightly turned in the direction of the orifice. At this moment there is also some tension on the fibres.

c   As the hand moves back towards the bottom of the dressed flax the wrist is turned away from the orifice and the thumb rolls along the index finger. The tension on the yarn is therefore held between the thumb and index finger and the twist is released into the forming thread. At the same time the fingers and thumb slide up the thread, smoothing any fibre ends into the twist.

d   As the hand returns there is no longer any tension with the fibres, so when the process is repeated the thumb rolls once more to the tip of the index finger, and as this is against the twist it is possible to select just sufficient fibres required for the next draft, while the tension on the yarn is transferred back to the third finger.

e   Only the left hand is used for both drafting the fibres and controlling the twist, but if too many fibres come down with the drafting (which is inclined to happen when starting a newly dressed distaff) the right hand can cross above the left hand and hold them back. When the left hand fingers need re-dampening, transfer the thread to the right hand after the fibres have been drafted and before smoothing; either hold the twist or smooth the thread with the right hand, by which time the left hand can take over the next drafting. There is no need to stop the steady treadling.

f   As the spinning progresses the distaff should be turned so that it is evenly spun off and the ribbon re-adjusted from time to time.

Method II
a   The fibres are drafted and controlled with the left hand while the right hand is kept damp to roll and smooth the thread. Thus the left hand movement is down-wards towards the orifice and the right hand upwards towards the distaff.

b   The fingers are re-dampened as the left hand drafts, so for this method the water pot needs to be close to the orifice.

c   If too many fibres come down with the drafting, the left hand holds them back and, to avoid stopping, until this is corrected the right hand both drafts and smooths.

Difficulties
a   Losing the end: this can happen with all fibres, but particularly with linen. To help avoid this, move the thread, not to the adjacent hook on the flyer, but to two or three away so that the thread is constantly getting crossed in the winding-on. When moving the thread from the hook nearest the orifice, take it to the hook at the far end of the flyer arm so that there is thread diagonally across the length of the bobbin. Should it be necessary to cut some away, if the end is irretrievably lost, it need only be as far as the last diagonal. The lost thread can sometimes be made to appear – except on spinning wheels that have a bobbin drag – by treadling fast in the opposite direction from when spinning. A small slip of paper laid lengthwise over the spun yarn on the bobbin from time to time can also help in finding a lost thread. A fine linen thread may also break during the reeling off when it is equally difficult to find the end unless the suggested precautions are taken.

b Too many fibres constantly coming down: probably the result of a badly dressed distaff. The fibres must be very finely separated, which takes time and care. Badly dressed or poor quality flax can be difficult to spread, which hampers fine spinning.

c The thread becoming too fine and disintegrating: possibly the driving band is too tight and the drawing-in too fast, or the hand is working too slowly.

d The twist running into the fibres on the distaff: lack of control of the twist, usually from failing to keep a tension on the spun thread. This can also be the cause of c.

e If difficulty is found in making the fibres stick together whilst spinning a fine thread, carrageen moss (obtainable at some health food stores) can be boiled in water and cooled to make a light glue-like fluid for wetting the fingers.

f Breaks: if the thread breaks (from any of the above reasons or from either too much twist or too little), remove any half formed thread or loose fibres hanging down from the distaff. Unwind a length of yarn from the bobbin, turning it by hand, and join as described in Spinning, *Method Ia*.

Distaffs dressed with the fibres straight down

This method of spinning can also be used with comb distaffs where the fibres have been 'kinked' (see p. 100, chapter IV). The distaff should be placed a certain distance away, usually above the orifice, to allow the spinner enough space for the drafting. The left hand holds a small bunch of the fibres at the bottom, and the right hand drafts a few at a time, pulling them out about 15in (38cm.) or so (according to the length of the fibres). The right hand only, or both hands, can smooth the thread on the return to the bottom of the fibres, keeping a constant tension on the thread.

Tow

All waste and short ends of flax can be placed in a tow distaff and spun from the mass. With this distaff it is necessary for one hand to hold the fibres back while the other does the drafting and smoothing; the draw is very short (see plate 35, chapter IV). If no tow distaff is available, hackle the tow and spin from the lap in a worsted manner; very short tow can be carded.*

PLYING

Plying gives a uniformity to the yarn and adds to its strength. It is usual, but not

*A tow distaff can easily be made from a branch of a tree with four or five smaller branches splaying out at one end. Be sure that the central branch is at least 4ft. (122cm.) in length and straight. It can be set in a hole in the centre of a stool or tied to a chair; if it will fit into a distaff arm on the spinning wheel, the stick can be shorter.

A distaff for line can be made from a branch which has four supple off-shoots which can be tied together to form a pear-shaped cage. If the wood is rough, cover with a piece of cloth (such as unbleached calico) cut on the bias and pulled tightly round the 'cage'. *Alternatively*, make a cone of thick brown paper approximately 15–18in. in depth and 7in. across at the bottom. Place on the top of a broom handle, and pack the hollow of the cone with tissue paper or newspaper to make it quite firm.

93 Plying

always necessary, to ply wool that is to be used for knitting or crochet, since if this is not done there is a tendency for the finished article to slope in the direction of the twist. Some warp yarns are better plyed, particularly if they have been spun in the worsted-type method and rather lightly. It is not so usual to ply linen, being a strong yarn, though it is a useful way of making neat thicker yarn.

For normal plying the direction of the twist is opposite to the twist of the singles yarns. Z singles are plyed in an S direction: S singles are plyed in a Z direction. If the yarns are plyed in the same direction as they were spun, the plyed yarn will become very over-twisted.

Check that the plyed yarn will not be too thick to go through the orifice of your spinning wheel. With this reservation, any number of yarns can be plyed together.

Place the bobbins of yarn to be plyed on a Lazy Kate or spool rack.* The bobbins are either placed side by side or one above the other, and they must be free to rotate, the yarns coming from underneath as they unwind. Place the rack on the floor on your right side (or whichever side you draw-out).

Alternatively: wind the yarn into hanks and place these on a double wool rice or two swifts, rotating and unwinding the yarns in the same direction. This will assist in keeping a uniform tension.

The starter thread on the bobbin must be twisted in the direction of the ply. Thread the yarns to be plyed through the orifice and eye and tie to the starter thread between hook and bobbin, so that the knot does not catch on the former.

*This can easily be improvised from a cardboard box and a knitting needle and is much more satisfactory than plying from balls of yarn in buckets.

The singles yarns can be kept separated by passing each between a different pair of right hand fingers, but the lengths being plyed should be held between index finger and thumb of the same hand. The long draw method allows approximately yard (91cm.) lengths to be plyed each time and makes for even distribution of twist. It can be rather tiring on the right arm and although it is possible to do this method with one hand only, it is best to hold with the left hand at the last point of twist as each new length is pulled from the bobbins. There should be absolutely even tension on the lengths of yarn as they are twisted together (93).

Alternatively: Some spinners prefer to let the yarns twist and be drawn-in continuously which is less tiring. It needs to be done at a regular speed in order to achieve uniform twist. The evenness of the tension on the yarns is again of utmost importance. It can be advantageous to pass the yarns through an eyelet, screwed into a bar placed above the bobbins, which keeps the yarns together and the bobbins unwinding evenly, making it easier to control the tension.

It will be noted that some of the twist will be removed from the singles yarns, so although the plyed yarn is thicker it is also softer. This should be considered when spinning singles, and if a more tightly spun and plyed yarn is required, the singles themselves should be given more twist.

In a well-balanced ply the angle of the twist should be about 45° and when worsted or semi-worsted yarns are used the fibres themselves should lie approximately parallel with the axis of the yarn.

If a lightly plyed yarn is required the yarns must be plyed at a faster speed, which might necessitate tightening the driving band to increase the draw-in. This will of course remove less twist from the singles.

If two plyed yarns need to be plyed together (sometimes called cabling) it can be done in either direction according to the amount of twist already in the yarns and the amount of twist required in the cabled yarn. Plying in the opposite direction to the first ply will make it softer.

REELING

Removing the yarn from the bobbin is usually done today by the spinner herself, so there is no need to remove the bobbin from the spinning wheel.

When the bobbin is full (or the yarn on it required) slacken the tension fully on the driving band and slip it off the bobbin whorl so that the bobbin can rotate freely.

Guide the thread round a part of the spinning wheel (such as the tensioner, the top of the maiden or a wheel upright) so that there is tension on the yarn and it does not catch on the hooks of the flyer. Always leave the stem of the bobbin thinly covered with yarn from which you can draw your next starter thread.

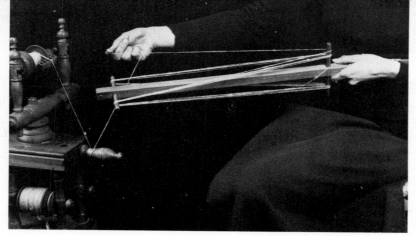

94 Reeling off, using a niddy-noddy which has the centre post extended to form a handle. The beginning of the yarn is wound round the handle. (On a niddy-noddy without a handle the beginning of the yarn is tied to one of the cross-bars.)

Form into a skein or hank, either by using a niddy-noddy (**94**) or reel, or if without equipment, round the palm of the hand and elbow.

Wool: finish reeling so that the beginning and end of the yarn meet with sufficient over-lap for tying them together with a single knot leaving at least 2in. (5cm.) of ends so that these can be found after scouring.

The hanks should be tied in four places very loosely with white thread (plyed wool or thick cotton is best) in a figure of eight, and the two ends of the thread knotted (*figure* 15 i).

Linen: it is better if the two ends do not meet round the skein but are joined by each end being tied to a length of strong white cotton. This makes the ends of the skein clearly visible after scouring.

Linen needs more ties and slightly firmer ones, since the yarn gets so easily tangled. To do this during reeling, have several lengths of white cotton (not wool since it might disintegrate in the scouring) and tie a half-knot (or half-hitch – *figure* 15 inset) lightly after a pre-determined number of turns (or with a click

i

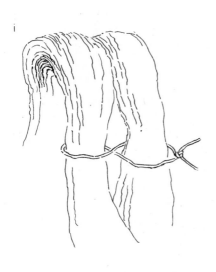

ii

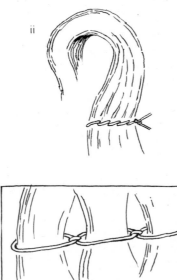

i. figure of eight tie on a hank of wool

ii. ties for linen: inset, detail of a half-knot

reel, after every click). This keeps a check on the yardage, and six or eight of these ties plus some figure of eight ties will keep the skein firm (*figure* 15 ii).*

N.B. Hanks of yarn that have the ties too tight will have patches remaining greasy or gummy after scouring. Never use coloured yarn for the ties, since the dye may run in the scouring.

SCOURING AND DRYING

Wool
It is usual to scour hanks of wool before use.

Wool yarns, woollen or worsted should be scoured in the same way as described for fleece (p. 194). If washed prior to spinning it is only the oil that is being removed at this stage, so a preliminary soaking should not be necessary.

Worsted yarn should be dried under tension; though this removes some of its elasticity it also reduces further shrinkage. Place the hanks, which must all be the same length, on to a dowel and sling between some sort of frame, such as a clothes horse. Place through the bottom of the hanks a second dowel on which weights can be hung.

Alternatively: small quantities can be re-reeled (not forced back on to the reel) to dry.

Woollen yarn can also be dried under tension to remove the crinkled look of hand-spun. The weights can rest partly on the ground, so that the hanks are stretched out but the springiness is not removed.

Linen
Linen yarn can be woven untreated (this is usual if it has been spun thick and plyed for rug warps), or softened, or bleached, though it is more usual to bleach woven cloth. For crochet it needs to be softened before use, but for lace it can be used untreated or softened, depending on the type of pattern. It is easier to ply fine linen if it has been softened first.

To soften: boil the hanks in water to which has been added $\frac{1}{2}$ of their dry weight in soapflakes, for about 1 hour. Unlike wool, linen can be moved about in the water and needs to be pressed down from time to time to keep it totally submerged. (The addition of a little washing soda can help de-gum.)

Rinse and squeeze out.

Boil the hanks for a further $\frac{1}{2}$ hour in fresh water, adding $\frac{1}{4}$ the dry weight in soapflakes and a little soda. Rinse, or if the thread is to be used for weaving, the soap can be left in after the last wash. This acts as a form of size, and will wash out when woven.

*If using a reel for linen a smoother, more polished thread can be achieved by passing the yarn through a wetted sponge held in one hand. This will smooth any fibre ends into the yarn and if a break has occurred, help to conceal the knot.

Hang out to dry in the open.

When dry there is a tendency for the threads to cling together and therefore, before use, it may be advisable to re-reel or wind on to spools.

Half-bleaching: the above method of softening will lighten the colour of the yarn and constant washing will eventually bleach it. To produce a creamy colour easily, weak solutions of proprietary brands of bleach can be used quite safely. Soften the yarn first as described above, then soak in a solution of cold water and bleach ($\frac{1}{4}$ tsp. of bleach to 1 pint – ·568 litre of water or as recommended on the bottle) for several hours. Rinse thoroughly and hang up to dry in the open. It may be necessary to repeat this process more than once, particularly if it was dark flax in the first place.

NOTES ON USING THE SPINDLE WHEEL

Some spinners like to use a spindle wheel for making woollen yarns and it is also useful for cotton. Some procedures have been described earlier in the book. From the practical viewpoint, particularly for those accustomed to using both hands on the flyer spinning wheel, the problem is to control the fibres and the twist with one hand only, while drawing-out away from the point of twist and making the yarn even.

Joining

The method of joining is not unlike that described for the flyer spinning wheel on p. 198 except that the yarn from the spindle tip is held between the thumb and the index finger of the left hand and the rolag, held in the palm of the hand, is on top (**95**). Do not attempt joining too near the spindle tip, and build up some twist on the starter thread. As soon as the wheel is turned with the right hand and

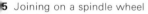

95 Joining on a spindle wheel

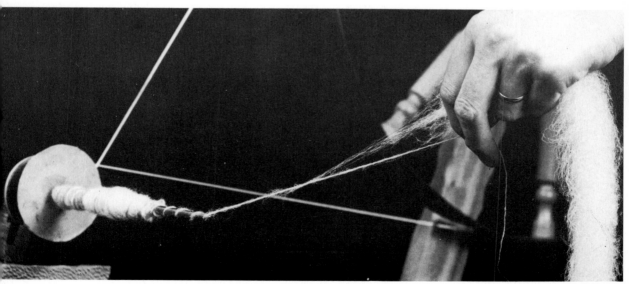

some fibres from the rolag start attaching themselves to the yarn, the left hand must immediately move away from the spindle before too much twist locks the fibres and makes it impossible to attenuate them.

Spinning

After each wind-on allow enough yarn to spiral up to the spindle tip and about 6in. (15cm.) beyond, and start drawing-out immediately (some say start drawing before the yarn has reached the spindle tip). Either the long draw method or, if the fibres are very short, the cotton draw method can be used. When using the long draw, it may be found easier (to start with anyway) to attenuate two or three shorter lengths of the rolag before the full arm span is reached, instead of only one as described for use on the flyer wheel, since a great wheel or wool wheel gives a much longer drawing-out distance. The relation between draw-out and twist must be carefully watched and if there is sign of the thread disintegrating, stop the left hand and build up some twist with the wheel. With most beginners the difficulty comes through the left hand not moving away fast enough, resulting in a 'rope' of yarn, rather than a thread. If too much tension is put on the spindle it will cease to turn, and moreover some old wheels with wooden spindles can stand very little. Allow some extra twist to each length of spun yarn before backing-off and winding-on. To ensure that the amount of twist is always the same, one can count the number of times the wheel is turned.

Building the cop or cone, and reeling

Start to build the cone or cop on to the spindle near to the front bearing. Always lay the yarn very close as it is wound on towards the spindle tip, and in wide spirals as it is guided back to the base. When the cone or cop is completed it can be eased off the spindle from the bottom and placed on a vertical spike for reeling. A well-built cone should unwind smoothly from on top.

Alternatively: slacken the driving band and reel direct from the spindle, *or*, before commencing spinning, fix a cardboard tube on to the spindle and build a cop on to this. It can then be reeled off horizontally from a rack or Lazy Kate, or put aside for plying.

Plying

No restriction on the thickness of the yarn is the reason for some people using a spindle wheel. For plying, however it is particularly useful. Use the long draw method (p. 218) as described for the flyer spinning wheel, but as the plyed yarn is backed-off and wound-on, more yarn is unwound from each of the cops. To ensure an even tension on the yarns, smooth from spindle tip between the left hand fingers at the start of each new twist.

NOTES ON USING THE SPINDLE

The spindle is a useful stepping stone to the spinning wheel. One can learn the 'feel' of spinning, giving all one's attention to the hands without the further co-ordination required in using the treadle.

All methods of spinning already described can be used on a spindle, but the short draw is the most usual. The hand movements are the same, but the hands work one above the other rather than beside each other.

To use a spindle that is suspended in the air, the length of spun yarn tied to the stick should be wound round a few times close to the whorl in the direction of the twist, spiralled to the tip of the stick, and held in place with a half hitch; the end of the thread comes up from under the last spiral. With a stick that has a hook or notch, the half hitch is slipped into this (see *figure* 2, chapter II).

Alternatively, the thread can first go under the whorl, round the tip of the stick and straight to the half-hitch. If the whorl and hook are above the stick the yarn is simply wound round the hook two or three times. Allow at least 8in. (20·5cm.) of yarn above the hitch.

The hand that puts the spindle in motion, whether by twirling or rolling on the hip (usually the right hip), is also going to control the twist. Hold the yarn with either a new rolag if joining, or a partly spun one at the last point of twist with the left hand. With the right hand give the spindle a sharp turn (to the right for Z and the left for S) and immediately it is set in motion the right hand takes over the position of the left hand, which starts to draw-out the fibres.

Few people today use a distaff with a spindle, so care has to be taken that the unspun yarn does not get caught in the twist. Lay the rolag along the back of the hand.

The usual difficulty lies in forming a thread strong enough to support the weight of the spindle; first attempts are best done over a soft surface (such as a rug) or with the tip of the spindle balanced on a hard surface or dish (see p. 41).

Keep an eye on the spindle to watch that it neither stops nor reverses its direction, which would undo all your work.

Do not let go the last point of twist while unhitching and winding on, since the same thing will happen.

SOME VARIATIONS IN CARDING

Blending qualities
Either mix thoroughly in the teasing, or keep separate and put equal quantities of each quality on to the carder in two thin layers each composed of half and half.

Blending fibres
This is a useful way of strengthening a weak fibre, enlivening a dull one, giving a fluffy appearance to a smooth one. Obvious successful mixtures are silk with cashmere, angora or wool; wool with mohair or angora. Fibre lengths should be matched as near as possible. It should be borne in mind that different fibres shrink differently and react differently in the dye-pot; these very differences can sometimes be used to effect.

Blending colours

Either put a thin layer of each colour on the carder, or card thin layers of each colour separately, placing them together for the final roll. Various shades of grey can be made from a natural dark brown and white according to the proportions used.

To keep colours separate on carders

The number of colour blocks must be uneven otherwise the colours will blend. If alternate blocks are required, make the first rolag by placing colour 1 on the two outer edges of the carder and colour 2 in the centre. Make the second rolag by placing colour 2 on the outer edges and colour 1 in the centre. Working on the same principle, smaller blocks can be divided into 5 or 7, and more colours can be introduced. When spun and knitted or woven the result can look as though a dip-dyed yarn had been used.

Bobbly effect

Little 'pills' of fibres can be kept and carded in with wool.

SOME VARIATIONS IN SPINNING WOOL

Spinning a fine yarn

In the first place use less wool and make thin rolags. Use the long draw method and draw-out quickly into a lightly twisted roving. Drop the rolag and draw-out between the hand small sections of the roving successively into a fine yarn.

Spinning a thick yarn

Use more wool and make large bulky rolags, for which the drum carder can be useful. Have a tight driving band so that the yarn is pulled in quickly; obviously the orifice must be large. Use the long draw method with hands working fast, since it is easy to over-twist.

Controlling long fibres

The spinner may find it easier to spin from the fibres folded over the index finger of one hand and with the other hand draft the fibres from the centre of the fold. This method can be used for long wool and flax (see descriptions on pages 183 and 191).

Spinning controlled slubs

To go through the orifice these must not be too big. In worsted spinning the slub is made by holding back small sections of wool with the front hand as the twist is allowed to run into the drafted thread. In woollen spinning the left hand jumps over small sections of unspun fibre as the right hand draws back. Care must be taken not to over-spin after the slub is made.

Making a marled yarn

Make two rolags of different colours. Draw out both together to form one yarn; it will be found that one will draw-out faster than the other, giving the uneven effect. If the two are controlled absolutely evenly the effect will be of a plyed yarn of two colours but in a singles form.

SOME VARIATIONS IN PLYING

Plying yarns under different tensions
To do this successfully it is necessary to hold a yarn in each hand and, to avoid entanglement, to have a Lazy Kate on either side of you (96). If one yarn is held tight the other will spiral round it. Nop yarn can be made by guiding one yarn back and forth over and over one part of the other yarn. If two colours are used the nop can be made with alternate hands.

Plying yarns of different twist
If an S twist and a Z twist are plyed Z, the Z will become over-twisted and the S under-twisted. If the S was lightly twisted in the first place a scalloped type of thread can be made. Commercial yarns can be plyed with hand-spun ones. The variations in thickness, directions of twist and variations in tensions are endless.

Blending
Plying two yarns from different fibres is a form of blending them. If two contrasting colours are plyed together the resulting yarn is speckled, but if the two colours are close in tone the resulting yarn is of intensified colour.

Insertions
Pieces of unspun fibre can be inserted in the twist; they should be prepared and be readily to hand to minimise disturbance of the rhythm of the plying.

A form of three ply
A Navajo technique of plying from one singles yarn is to make large interlinked loops by drawing them through with the finger and then letting them receive twist. It is similar to chain stitch in crochet but the loops are kept very large, making it easy to make the subsequent loops with the finger. This is particularly useful when a singles thread is spun in blocks of colour which one wishes to retain (*figure* 16).

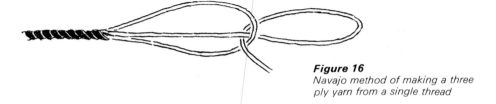

Figure 16
Navajo method of making a three ply yarn from a single thread

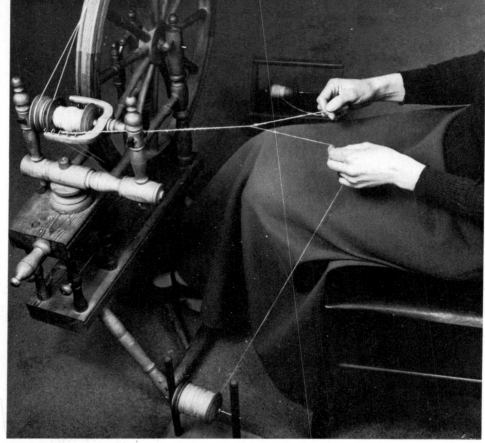

96 Plying, with independent control on the tension of a yarn

DESIGN AND PURPOSE

Variations in designing a thread are limitless but the choice must rest with the yarn's suitability for its final use. At the same time the article envisaged should warrant the time that it takes to spin yarn. Samples of the yarn, the yarn into fabric, and the fabric 'finished' should be made to ensure that the fibre is suitable throughout and that the methods used are right for the fibre. Through such samples, knowledge and experience are built up.

When choosing a fibre, consider its feel, whether smooth or rough, coarse or fine, soft or hard and its amount of suppleness. Also its strength, its decorative value, amount of lustre, washing potential (if relevant), amount of elasticity, and amount of warmth (if concerned).

Spin a sample, perhaps making variations in thickness, amount of twist and possibly method of spinning; bear in mind the loss of twist if yarn is to be plyed. A yarn cannot be assessed when under tension or on the bobbin. When you have decided on the type of yarn you need, keep a sample beside you. It is so easy to forget details by the time you have filled even one bobbin let alone several. To allow for shrinkage, quantities should be calculated after scouring.

Hand spinning is but the beginning of a textile creation. In itself it gives pleasure and satisfaction, but the complete sense of achievement comes with the making of an article, large or small, perhaps useful, perhaps not, but pleasing both to the touch and to the eye.

97 Fibres and yarns from left to right:
i. Staple of Wensleydale (long wool and lustre) unscoured and not teased, ii. Worsted yarn plyed, iii. Worsted yarn singles, iv. Staple of Herdwick (mountain and hill) teased, v. Three ply rug yarn from Herdwick, vi. Finer yarn from the same wool, vii. Staple of Clun Forest (down and short wool) scoured and slightly teased, viii. Over-twisted yarn, ix. Under-twisted yarn, x. Woollen spun yarn, xi. The same yarn plyed, xii. Lightly twisted slub yarn plyed with S twist fine yarn, plyed S, xiii. Brown and white knop yarn, xiv. Marled and speckled effect from spinning a brown and a white rolag at the same time

98 Fibres and yarns from left to right:
i. Mohair; yarn and staple, ii. Angora rabbit; fine plyed yarn, thicker fluffier yarn and staple, iii. Samoyed dog hair; plyed, singles and some hair, iv. Silk; yarn and a portion of tops, v. Cotton; yarn and rolag, vi. Flax; fine linen, thicker and rougher thread and a finger of flax

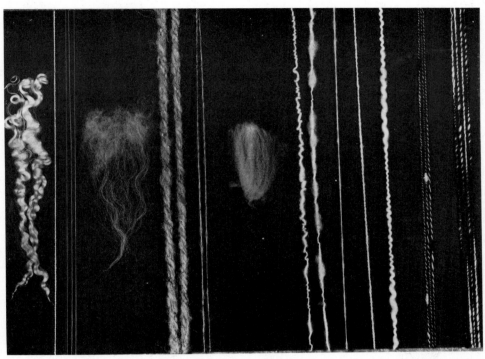

100 Two samples of lace made by Margaret Charlett with hand-spun linen:
(a) torchon insertion with the yarn softened and half-bleached; (b) $\frac{1}{4}$in. (1cm.) mesh braided lace with tallies, with untreated linen yarn

99 Detail of a woollen coat fabric; the weft is made with white slubs inserted at intervals into a brown yarn

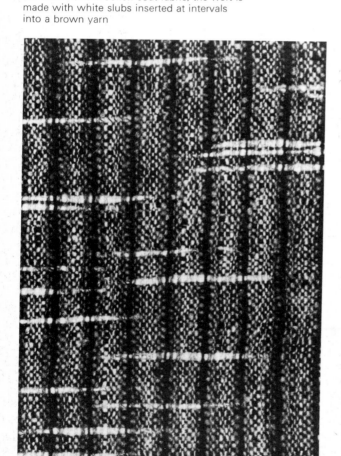

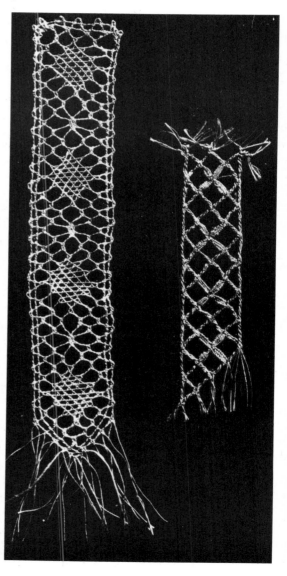

Appendix One

Choosing and sorting wool

CHOOSING AND SORTING WOOL

The fibre most used by hand-spinners today is wool. It also happens to be the easiest for the beginner and is perhaps the most versatile.

Although there are over 40 registered sheep breeds in Britain, over 400 different grades of wool are listed in the British Wool Marketing Board's schedule for the trade, each grade having its own number, and in the sales wool is bought and sold by the trade numbers, not by the name of the breed. There is also continual cross-breeding, and the expression 'comeback' refers to crossing and re-crossing to a return to the original breed. To protect some of the older breeds from becoming extinct there is a Rare Breeds Survival Trust. Wool from some of these breeds can be of interest to the hand-spinner.

The bulk of wool comes from the merino and British sheep breeds, now all over the world, and although classification varies in different countries, in England anyway, the hand-spinner still gains her experience from knowing the breeds along with a basic knowledge of how to sort out qualities of wool on a fleece.

Normally a sheep is sheared for the first time after it is at least 12 months old, when long-woolled sheep are often known as hogs and short-woolled sheep are called tegs. The first clip can be recognised by the tips of the staples being more pointed and fine. Subsequent shearings (when sheep are called wethers) have more rounded tips. Rams usually produce heavier fleeces than ewes.

Some points to look for when choosing a fleece

Condition of fleece:
Choose a clean fleece. Dirt can add considerably to the weight and therefore the price. Straw and vegetable matter can be tiresome to remove. Tar cannot be removed. (Farmer's markings should be water soluble.)

Choose a fleece with a sound fibre. To test, take a staple of wool and pull it hard from both ends between the hands. If it has a weakness it will break or show signs of disintegrating.

Avoid a fleece which is cotted, i.e. matted at the clipped end. It can be difficult to separate and if really bad is not suitable for the hand-spinner.

Avoid a fleece that has second clip ends. This occurs when the sheep was sheared too late and the short new growth gets clipped as well.

Look for dark fibres in the wool, which could spoil a white cloth and might also show when dyed.

Look for kemp, the coarse hair in the wool, sometimes different in colour, which will not normally take the dye. However wool that is kempy can make effective yarns for certain uses.

For suitability of purpose consider:

Fineness of fibre.

Handle (i.e. the feel), whether soft, harsh, tender, strong, springy, wiry, silky.

Length of fibre – look for uniformity.

Density of staple; fibres that grow close together and are fine will usually felt well.

Amount of crimp.

Amount of lustre.

SORTING

Purpose

To match the qualities of wool to obtain an overall evenness in the finished product. Different qualities can behave in different ways during the successive processes of fleece to fabric (e.g. shrinkage, felting, take-up of dye etc.) and, of course, can give different feels.

Coloured fleeces, when there is more than one shade, may be sorted for colour; even so, qualities should be taken into account.

Large quantities of matchings may require wool from several fleeces. Bear in mind that weight-loss after washing can amount to approximately a third.

Method

Unwind the neck rope, unroll the fleece and lay out, tips uppermost, shaking out the sections as you go, so that loose short ends of wool and dirt will drop out. Open up the neck wool. As far as possible, push the fleece into a shape as shown in *figure* 17.

Remove dags and dirt from around the tail and belly, known as skirting, if this has not already been done. Always push the wool gently apart with the fingers of both hands working away from each other. Avoid any unnecessary tearing at the wool and do not cut it.

Take sample staples from the various parts of the fleece and compare them. Take time and feel all over the fleece, and when you have decided where the quality changes, gently make the breaks.

When parted into sections, match up where and if possible.

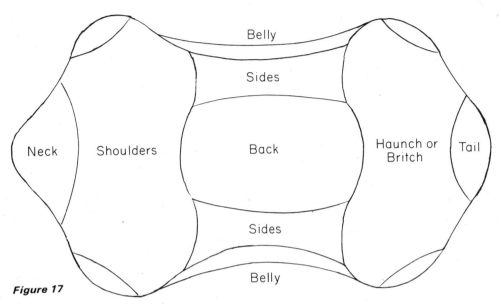

Figure 17

*Diagram of the layout of a fleece
ready for sorting*

Qualities

Shoulders are normally the best quality.

Sides may match or may be less good.

Neck and belly can have a soft handle but the latter is often too dirty for use.

Britch or haunch and any wool from the upper part of the legs (positions indicated but not named in *figure* 17) is usually much coarser.

Back, which takes the brunt of the weather, can be very dry, open and tender.

Storing

Label carefully noting the breed, date and quality.

Store in sacks if possible, with layers of newspaper between each quality, or use separate paper bags. Plastic bags, if used, should be well punctured to prevent undue sweating.

Appendix Two

SHEEP BREEDS

Breed	Quality count	Average staple length	Fleece weight	Comments
Merino White face & feet; curly horns.	65s–100s, many different grades.	2–4in. (5–10cm.)	9lbs. (4·08kg.) upwards	An old breed which probably originated in Spain, now raised in many parts of the world, particularly Australia. Wool has very soft handle and is crimpy. As tips are usually dirty and greasy it often needs careful scouring before spinning. Shrinks & matts easily. Can be spun very fine.

Downs & Short Wools

If the handle is soft many of the wools are suitable for making shawls, knitted garments, and the very best quality for babies' garments. The stronger wools can be used for light tweeds, travelling rugs, cushion covers etc., and spun into a bulky yarn for blankets.

Breed	Quality count	Average staple length	Fleece weight	Comments
Clun Forest Dark brown face & legs; forelock of white wool.	56s–58s	3–4in. (7·6–10cm.)	5–7lbs. (2·26– 3·17kg.)	Possibly an off-shoot of the Ryeland breed. Dense and fine wool. Very suitable for woollen spinning and for the beginner.
Devon Closewool Thickset; head & legs well covered with wool.	46s–50s	3–4in. (7·6–10cm.)	5–7lbs. (2·26– 3·17kg.)	Originated from a cross between Devon Longwool & Exmoor Horn.
Dorset Down Medium size; hornless with grey-brown nose & ears; wool on legs.	56s–58s	2–3in. (5–7·6cm.)	5–7lbs. (2·26– 3·17kg.)	Probably developed from crossing local sheep with Hampshire & Southdown. Has full springy handle.
Dorset Horn Curly horns; white face & legs; wool forelock.	54s–58s	3–4in. (7·6–10cm.)	5–7lbs. (2·26– 3·17kg.)	An ancient breed mentioned earliest in 17th century Polled Dorset (i.e. hornless) derived by crossing with Ryeland; curly horns interfere with shearing. Good wool for beginners; has excellent felting qualities and can be used for fine tweeds.

Hampshire Down Dark brown nose & legs; hornless. Woolled on forelock, cheeks & legs.	56s–58s	2–3½in. (5–9cm.)	5–6lbs. (2·26–2·72kg.)	First flock book 1890; a cross of local sheep with Southdown and Wiltshire Horn. Fine with rather short staple.
Kerry Hill Black & white woolless legs & face; black nose, pointed ears & a forelock.	56s–58s	4in. (10cm.)	5lbs. (2·26kg.)	Probably of same origin as Welsh Mountain but different environment and crossed with Clun Forest. Handle a little harsher than some of the Down breeds. Good for beginners, useful for knitting.
Llanwenog Black face, ears & legs; woolly forelock, hornless.	56s–58s	3in. (7·6cm.)	5lbs. (2·26kg.)	Developed at end of last century from crosses between local black sheep of Llanllwni & Shropshire. Good for beginners; and useful for knitting, cushions and clothes.
Oxford Broad with dark face; well covered with white wool.	50s–58s	6in. (15cm.)	8–10lbs. (3·62–4·53kg.)	19th century cross between Hampshire, Cotswold and a little Southdown. Often used for cross-breeding. Useful for lightweight clothing.
Portland Small with white face.		3–4in. (7·6–10cm.)	1½–3lbs. (·680–1·36kg.)	Ancestor of the Dorset Horn; protected breed, flock now being established on Portland Bill. Creamy white fleece with soft handle. Good for beginner; useful for soft weft yarns and knitting.
Radnor Grey nose; tan face & legs.	50s–58s	3–7in. (7·6–18cm.)	4–5lbs. (1·81–2·26kg.)	Probably a mixture of Welsh mountain, Clun Forest and Shropshire.
Ryeland Hornless; wool on white face & forelock.	56s–58s	3–4in. (7·6–10cm.)	6–8lbs. (2·72–3·62kg.)	From rye-growing areas of South Herefordshire since 16th century when known as 'Lemster ore'. Springy dense wool.
Shropshire Medium size with black face & legs; wool on cheeks & forelock.	50s–58s	3½–4½in. (9–11·5cm.)	6–8lbs. (2·72–3·62kg.)	A breed of this kind probably in 16th century. Good quality fine & dense wool; suitable for the beginner.
Southdown Compact build with face covered with short wool; hornless.	56s–60s	1–2½in. (2·5–6·5cm.)	6lbs. (2·72kg.)	Oldest of the English Down breeds, from which most others were developed late in 18th century by John Ellman of Glynde. Can be difficult for the hand-spinner because of short, fine staple.

	Quality count	Average staple length	Fleece weight	Comments
Suffolk Black face, ears & legs. Ruff round neck.	54s–58s	2–3in. (5–7·6cm.)	5½–7lbs. (2·49–3·17kg.)	Cross between Norfolk ewes & Southdown rams. Often used for cross breeding and adds soft handle to many of the breeds it is mixed with. Usually a nice wool for hand-spinners.

Three Scottish Island Breeds

Breed	Quality count	Average staple length	Fleece weight	Comments
Orkney Small with short tail. Rams horned; ewes hornless.	50s–56s	1½–3in. (4–7·6cm.)	2½–5lbs. (1·13–2·26kg.)	Represents ancestral stock from which Shetland developed. Now remains on North Ronaldsay, most N. of Orkneys. Lives on seaweed. Colouring similar to Shetland. Tendency to shed fleece. Very kempy.
Shetland Small with bright eyes.	56s–60s	4in. (10cm.)	2–3lbs. (·906–1·36kg.)	Possibly Scandinavian origin. Various colours, moorit (moor red), fawn, brown, greys and white which is usually softest. Used to be rooed (i.e. plucked). Only one or two pure flocks left. Now crossed and usually only 60% true Shetland. Shoulder wool can have remarkably soft and silky handle.
Soay Small; horned.	44s–50s	2–6in. (5–15cm.)	3–5lbs. (1·36–2·26kg.)	Close affinity with a wild sheep; two shades of brown with white belly. Lives on the islands of St Kilda.

Three Four-Horned Breeds

Breed	Quality count	Average staple length	Fleece weight	Comments
Jacob Brown & white patches.	44s–56s	3–6in. (7·6–15cm.)	4–6lbs. (1·81–2·72kg.)	Origin uncertain; Mesopotamia & Spain both suggested. Very popular with hand-spinners but quality variable because kept in so many areas. White usually softer than brown.
Manx Loughtan Fawn coloured.	44s–48s	3–5in. (7·6–13cm.)	4–6lbs. (1·81–2·72kg.)	Native to the Isle of Man. Loughtan means fawn-coloured. Little visible crimp, soft handle. Useful for woollen weft yarns.
St Kilda Small & black.	48s–50s	2–6in. (5–15cm.)	3–5lbs. (1·36–2·26kg.)	Appears to have originated in the Hebrides. Protected breed. Handle very variable; can feel quite silky but more often too harsh to be used for garments worn next to the skin.

Long Wools & Lustres

These wools are most suitable for worsted spinning, particularly if lustrous, since this type of spinning helps to retain the natural shine. Often coarse and strong, they are useful for hard-wearing suitings, upholstery etc.

Breed	Quality count	Average staple length	Fleece weight	Comments
Border Leicester Large, white with woolless face and 'Roman' nose; hornless.	40s—46s	6—10in. (15—25·5cm.)	6—10lbs. (2·72—4·53kg.)	18th century cross between English Leicester & Cheviot. If not too long a staple, can be good wool for beginners and if crossed with a Suffolk can be fine as well as lustrous.
Cotswold Large, hornless with forelock.	46s—48s	7—10in. (18—25·5cm.)	12—13lbs. (5·43—5·85kg.)	An old breed now amongst those protected and rare; similar to the Leicester with white curly fleece; inclined to cott.
Dartmoor Medium size, hornless; mottled white & grey face. Wool on head.	36s—40s	10—12in. (25·5—30·5cm.)	15—18lbs. (6·79—8·15kg.)	No clearly defined origin; might have Lincoln ancestry. Long curly fleece.
Devon Longwool Medium size; head well covered with curly wool.	32s—36s	12in. (30·5cm.)	14 lbs. (6·34kg.)	Obscure ancestry, but descendant of the Lincoln and improved by crossing with the Leicester. Long curly fleece. An open staple, lustrous and very strong; free from kemp.
Kent or Romney Marsh Large with white face and black nose; hornless and immune to foot rot.	48s—56s	6—7in. (15—18cm.)	8½—10lbs. (3·85—4·53kg.)	Its history on the Romney Marsh can be traced for 700 years. Amongst the most popular of the lustres with hand-spinners, and the shorter staple suitable for the beginner of worsted spinning.
Leicester Large & hornless with woolly forelock. Woolless white legs & face; dark nostrils.	40s—46s	12—14in. (30·5—35·5cm.)	11—13lbs. (4·98—5·85kg.)	An ancient breed believed to have been introduced by the Romans, but improved by Robert Bakewell of Dishley in 18th century. Wool has silky appearance & high lustre.
Lincoln Longwool Large, white face, hornless; woolled legs & head.	38s—44s	12—16in. (30·5—40·5cm.)	11—14lbs. (4·98—6·34kg.)	Probably the first of the English lustre breeds; not many crimps per inch.
South Devon Medium size with well-woolled head.	36s—40s	7in. (18cm.)	12—16lbs. (5·43—7·25kg.)	The wool is curly, lustrous and very dense.

Teeswater Medium size, hornless and grey or white head; forelock. No wool on face or legs which are sometimes black & white.	40s–48s	12in. (30·5cm.)	12–15lbs. (5·43– 6·79kg.)	For at least 150 years rams bred for crossing with various breeds; very curly wool.
Wensleydale Medium size with blue-grey face & woolled head.	40s–48s	14in. (35·5cm.)	10–16lbs. (4·53– 7·25kg.)	Breed established in 19th century by crossing a Yorkshire long wool breed with Leicester. Known locally as the Ripon hog; very long curly lustrous wool.
Whiteface Dartmoor Nowadays only the ram may have horns. Ewes have wool-free faces.	32s–44s	10–12in. (25·5– 30·5cm.)	10–14lbs. (4·53– 6·34kg.)	Known on the Moor since earliest records. White with fairly strong staple.

Mountain & Hill

Some of these wools can be very coarse and are useful for carpets, rugs and tough outer garments, but usually too scratchy to be close to the skin.

Breed	Quality count	Average staple length	Fleece weight	Comments
Cheviot Medium size; white woolless face & legs; ruff round neck.	50s–56s	4in. (10cm.)	4–5lbs. (1·81– 2·26kg.)	Lowland Scottish, probably originally descended from tan- faced moufflon. Between 15th and 17th century local Cheviot sheep crossed with merino and in 18th Century improved with Southdown. Of superior quality to most other mountain & hill breeds. Has demi-lustre and a strong fibre. Very popular with hand-spinners. Best quality can be spun worsted; spun woollen can be used for tweeds, knitting etc.
Dalesbred Woolless black face with white marks & curled horns.	32s–40s	8in. (20cm.)	4lbs. (1·81kg.)	A native of the Northern Pennine areas. Tough springy wool; curly on the outside & dense undercoat.
Derbyshire Gritstone Large with small head; black & white woolless legs.	50s–56s	6–8in. (15–20cm.)	5lbs. (2·26kg.)	An old breed of the Peak District.
Exmoor Horn Small & chubby white face and legs covered with wool; both sexes have horns.	50s–56s	3–4in. (7·6–10cm.)	5–7lbs. (2·26– 3·17kg.)	Old breed from West of England.

Herdwick Small with white woolless face; born almost black & gets lighter as it grows older. Rams have horns.	28s–32s	5in. (13cm.)	3–4½lbs. (1·36– 2·04kg.)	Old breed from the Lake District possibly introduced from Scandinavia. Has homing instinct. Wool prone to kemp but makes interesting yarn, particularly for rugs. Usually an easier wool for the hand-spinner than Scottish Blackface.
Lonk Large & strong. Black & white woolless face & legs; horns.	32s–44s	4–6in. (10–15cm.)	5–6lbs. (2·26– 2·72kg.)	Old breed from the Pennines of Yorks & Lancs; also has homing instinct. Used for crossing with Wensleydale, Suffolk, Leicester & Kerry Hill. Fine grades useful for knitted garments.
North Country Cheviot Larger than Cheviot; free from kemp.	50s–56s	2–3in. (5–7·6cm.)	3½–5lbs. (1·59– 2·26kg.)	Improved in 19th century in the Highlands of Scotland. Used for tweeds, overcoats, jackets & travelling rugs.
Rough Fell Black & white woolless face with horns.	32s–36s	8in. (20cm.)	5lbs. (2·26kg.)	An old breed from Westmorland; wool nearly touches ground.
Scottish Blackface Small; black & white woolless face & legs; curled horns.	28s–32s	8–13in. (20–33cm.)	4–7lbs. (1·81– 3·17kg.)	Origins unknown although prevalent in Scotland & northern England from 17th century. Very hardy breed. Has bottom wool (i.e. soft undercoat) and long hair which has little bounce unless mixed, when suitable for rugs, tough fabrics for upholstery & outer garments.
Swaledale Dark face & grey muzzle; both sexes have horns.	28s–32s	8–12in. (20–30·5cm.)	4lbs. (1·81kg.)	Old breed from the area which may have been cross between Scottish Blackface & Cheviot. Has long hair and soft bottom wool; lacks lustre.
Welsh Mountain Small with bright eyes; rams have horns. Slim woolless legs.	36s–50s	2–4in. (5–10cm.)	3lbs. (1·36kg.)	Careful breeding has improved this breed, and it usually has softer handle than most mountain breeds. Occasionally red-pink kemp found.
Black Welsh Mountain Small; very dark brown woolless face & legs. Rams have horns.	36s–50s	4in. (10cm.)	3lbs. (1·36kg.)	Has been bred selectively as a distinct breed for about 100 years. The short very dark brown wool is thick with a firm handle, but often wiry.

Whitefaced
Woodland
Horned with grey face.

Ancient breed from Peak District. In 18th century crossed with merino, now a protected breed.

Cross-Breds

All English breeds, because their quality is under 60s are known as cross-breds. Those above 60s are the merino types. No flock books date before the eighteenth century and sheep breeds as we know them today date from the nineteenth and twentieth centuries. Cross breeding and selective breeding is done continually, to improve quality and uniformity. A knowledge of the wool characteristics of the breeds provides a guide to the type of fleece that is likely to be found in the cross. Some cross-breds take on new names, and in New Zealand and Australia where the merino strain has been introduced to English breeds, new breeds have been established.

Some British cross-breds:

Name	Quality count	Average staple length	Fleece weight	Cross
Colebred Woolless white face & legs.	48s–54s	4–7in. (10–18cm.)	5–7lbs. (2·26– 3·17kg.)	Border Leicester x Clun Forest x Friesland (from Holland) x Dorset Horn.
Masham Medium; grey face & hornless. or can have black & white faces.	44s–48s	6–7in. (15–18cm.)	5–7lbs. (2·26– 3·17kg.)	Swaledale x Wensleydale or Rough Fell. Teeswater x Dalesbred or Swaledale or Rough Fell.
Romney (or Kent) Halfbred	56s–58s	2–4in. (5–10cm.)	6lbs. (2·72kg.)	North Country Cheviot ram x Romney ewe or Romney x Southdown.
Scottish Greyface Known as Greyface or Mule.	32s–48s	6–12in. (15–30·5cm.)	6–8lbs. (2·72– 3·62kg.)	Border Leicester ram x Scottish Blackface ewe.
Scottish Masham	—	—	—	Teeswater x Scottish Blackface.
Welsh Halfbred	48s–54s	5–6in. (13–15cm.)	6–7lbs. (2·72– 3·17kg.)	Border Leicester x Welsh Mountain.

Some New Zealand breeds which are now established as a consequence of cross-breeding:

Name	Quality count	Average staple length	Fleece weight	Comments
Coopworth	46s–50s	4–6in. (10–15cm.)	—	Established quite recently. Border Leicester x Romney.
Corriedale	50s–58s	6–8in. (15–20cm.)	10–12lbs. (4·53– 5·43kg.)	Established nearly 90 years. Selected breeding from Leicester & Lincoln half-breds (i.e. x Merino).
Drysdale	30s–34s	10–14in. (25·5– 35·5cm.)	—	Developed around 1945 for carpet trade. Hairy type of Romney x Romney/Cheviot ewes.
Perendale	46s–56s	5–7in. (13–18cm.)	8–9lbs. (3·62– 4·08kg.)	Established about 32 years. Romney x Cheviot.
Polwarth	58s–64s	3½–4½in. (9–11·5cm.)	—	Established over 90 years. Merino x Lincoln.

Source material

'Account of the machines and models collected by the Honourable Board of Trustees for manufactures for Scotland.' 1784. Manuscript in Scottish Record Office.

ADOLFS-UTSTÄLLNINGEN, G.
Redskapsstudier.
Stockholm, 1933

ALFORD, V.
Pyrenean Festivals.
London, 1937

ANDERSON, E.
Plants, Man and Life.
Berkeley, California, 1971

ANONYMOUS
Floral fancies and morals from flowers.
1843

ASPIN, C. & CHAPMAN, A. D.
James Hargreaves and the Spinning Jenny.
Helmshore local history society, 1964

AUBREY, J.
Natural History of Wiltshire.
1697, Manuscript in the Bodleian Library, Oxford

BAINES, E.
The History of the Cotton Manufacture.
London, 1835; reprint 1966

BAINES, E.
The Woollen manufacture in England (in T. Baines's *Yorkshire, Past and Present*, 1875); reprint Newton Abbot, 1970

BAUR-HEINHOLD, M.
Deutsche Bauernstuben.
Stuttgart, 1967

BENJAMIN, F. A.
The Ruskin Linen Industry of Keswick.
Cumbria, 1974

BERG, J. & LAQUERANTZ, B.
Scots in Sweden.
Stockholm, 1962

BINNS, J.
Miseries and Beauties of Ireland.
London, 1837

BODMER, A. M.
'Spinnen und Weben im Französischen und Deutschen Wallis'.
Romanica Helvetica Vol. 16.
Geneva, 1940

Book of Trades. (ed. Phillips, R.)
London, 1804 and 1815

BOMANN, W.
Baüerliches Hauswesen und Tagewerk im alten Niedersachsen.
1927

BRIDGE, A. & LOWNDES, S.
The Selective Traveller in Portugal.
London, 1967

BURNHAM, H. B. & D. K.
'Keep me warm one night', early handweaving in Eastern Canada.
Toronto, 1972

BURT, W.
Review of the Port of Plymouth.
Plymouth, 1814

CHANNING, M.
Textile Tools of Colonial Homes.
Marion, 1969

CIBA REVIEWS:
no. 10. LEIX, A.
Trade Routes and Dye markets in the Middle Ages.
1938

no. 14. GUTMANN, A. L.
Cloth making in Flanders.
1938

no. 20. WITTLIN, A.
The Development of the Textile Crafts in Spain.
1939

no. 27. REININGER, W.
The Textile Trades of Medieval Florence.
1939

no. 28. BORN, W.
The Spinning Wheel.
1939

no. 29. DE FRANCESCO, G.
Venetian Silks.
1940

no. 48. DRIESSON, L. A.
The History of the Textile Crafts in Holland.
1944

no. 59. SCHWARZ, A.
The Reel.
1947

no. 64. WESCHER, H.
Cotton and the Cotton Trade in the Middle Ages.
1948

no. 80. EDLEIDE DE ROOVER.
Lucchese Silks.
1950

no. 111. TRAUPEL, R.
Spun Silk.
1955

CLARK, A.
Working life of women in the 17th century.
London, 1919

COBBETT, W.
Rural Rides.
London, 1923

COOK, J. G.
Handbook of Textile Fibres. Vol. I. Natural Fibres.
Watford, 1959

COOTE, C.
Statistical Survey of County Monaghan.
Dublin, 1801

COWLIG, W. T.
It happened round Manchester. Textiles.
London, 1970

DANILOFF, S.
'Some rare spinning wheels'.
Antiques vol. 16.
1929

DEAN, I. F. M.
Scottish Spinning Schools.
London, 1930

DEFOE, D.
A tour through the whole Island of Great Britain.
London, 1724

DELAPORTE, L'ABBÉ Y.
Les Vitraux de la cathédrale de Chartres.
Vols. II & III.
Chartres, 1926

DELONEY, T.
The pleasant history of Jack o'Newbury.
1597

DICZIUNARI RUMANTSCH GRISCHUN.
Section 'Filadè I.'
Winterthur, 1974

DIDEROT & D'ALEMBERT
Encyclopédie
'Fil' Vol. VI 1756; 'Laine' Vol. IX 1765;
Plates 'Fil et Laine' Vol. XXI 1765. Paris

DUFF, D.
Victoria in the Highlands.
London, 1968

DYER, J.
The Fleece.
London, 1757

EDWARDS, R.
The Dictionary of English Furniture Vol. III
'Spinning wheels'.
London, 1924; revised 1954

Encyclopaedia Britannica 11th Ed.
Vol. XXV. Fox, T. W. 'Spinning'
Vol. X. Woodhouse, T. 'Flax'

Encyclopedia of Textiles. (American Fabrics Magazine)
Englewood Cliffs, N. J., 1960

Encylopédie. see DIDEROT

ELLACOTT, S. E.
Spinning and Weaving.
London, 1956

ENGLISH, W.
The Textile Industry.
London, 1969

ENDREI, W.
L'évolution des techniques du filage et du tissage.
Paris, 1968

FELDHAUS, F. M.
'Meister Jürgen'
Neue Allgemeine Deutsche Biographie Vol. 50.
1875–1912

FELDHAUS, F. M.
Die Technik der Vorzeit, der Geschichtlichen Zeit und der Naturvölker.
Leipzig, 1914

FELDHAUS, F. M.
Die Technik der Antike und des Mittelalters.
Potsdam, 1931

FIENNES, C. (ed. Morris, C.)
Journeys (c.1685–1703).
London, 1947; revised 1949

FIRMIN, T.
Some proposals for imployment of the poor. . .
London, 1681

FORBES, R. J.
Studies in Ancient Technology.
Vol. IV. Textiles.
Leiden, 1959

GAINES, R.
'Homespun'.
*The Chronicle of the Early American
Industries Ass.*
Vol. II. no. 17.
1941

GANDHI, MOHANADASA K.
The Wheel of Fortune.
Madras, 1922

GILBERT, C.
'John Planta of Fulneck, Yorkshire'
Furniture History. Vol. VI.
1970

GILBERT, K. R.
Textile Machinery.
London, 1971

GILL, C.
The Rise of the Irish Linen Industry.
Oxford, 1925

GLEMNITZ, J.
Wolle spinnen am Handspinnrad.

GLOAG, J.
A Short Dictionary of Furniture.
London, 1952

GRAHAM, H. G.
Social Life of Scotland in the 18th century.
London, 1937

GRANT, MRS. A.
*Essays on the superstitions of the Highlanders
of Scotland.*
1811

GRANT, I. F.
Highland Folk Ways.
London, 1961

GRAY, A.
A Treatise on Spinning Machinery.
Edinburgh, 1819

GREGORY, E. W.
The Furniture Collector.
London, 1927

GROVES, S.
The History of Needlework Tools.
London, 1966

GUEST, R.
*A Compendious History of the Cotton
Manufacture.*
Manchester, 1823

HABIB, I.
'Technological changes and society in the
13th and 14th centuries'.
Presidential Address, Medieval India
Section, 31st session of the Indian History
Congress, Varansi.
December, 1969

HALL, R.
*Observations . . . on the methods used in
Holland, in cultivating or raising of hemp
and flax.*
Dublin, 1724

HANCOCK, H. B.
'Furniture Craftsmen in Delaware Records'
Winterthur Portfolio IX.
1974

HANSEN, H. P.
Spind og Bind.
Copenhagen, 1947

HARTLEY, M. & INGLEBY, J.
The Old Handknitters of the Dales.
Yorkshire, 1951

HAVINDEN, M. A. (ed.)
*Household Farm inventories in Oxfordshire.
1550–1590.*
London, 1965

HENTSCHEL, K.
Wolle spinnen mit Herz und Hand.
1949; reprint 1975

HILLS, R. L.
Power in the Industrial Revolution.
Manchester, 1970

HOFFMANN, M.
'En hjulrokk sum ville konkurveve med
spinnemaskinen'
By og Bygd, bind 6.
Oslo, 1948–49

HOFFMANN, M.
'The Great Wheel in Scandinavian Countries'
Studies in Folklife.
1969

HORNER, J.
*The Linen Trade of Europe during the Spinning
Wheel Period.*
Belfast, 1920

HOWARD, J.
The State of the Prison in England and Wales.
Warrington, 1777

HUGHES, G. B.
'The Evolution of the Spinning Wheel'
Country Life, 1959

HUGHES, T.
Cottage Antiques.
London, 1967

JAMES, J.
History of the Worsted Manufacture in England from Earliest Times.
London, 1857

JENKINS, J. G.
The Welsh Woollen Industry.
Cardiff, 1969

KILLIP, I. M.
'Stiff carts and Spinning Wheels'
Journal of the Manx Museum Vol. VI. no. 77.
1960–61

KILLIP, I. M.
'Notes on Manx Homespun'
Journal of the Manx Museum Vol. VI. no. 81.
1965

KOLCHIN, B. A.
Arkheologiya SSSR E1–55 Moscow, 1968.

Kunstindustrimuseet & Herning Museum.
Spind & Tvind.
Copenhagen, 1975

LAWSON, J.
Progress of Pudsey during the last Sixty Years.
1887. (Part reprinted with Baines's *Account of the Woollen Manufacture of England,* 1970.)

LEMON, H.
'The hand-craftsman in the wool textile trade'
Folklife, 1963–64

LEMON, H.
'The Development of hand spinning wheels'
Textile History Vol. I.
1968

LEONARD, E. M.
The Early History of Poor Relief.
Cambridge, 1900

LINDER, A.
Spinnen und Weben einst und jetzt.
Lucerne, 1967

LING-ROTH, H.
'Hand Woolcombing'
Bankfield Museum notes.
Reprint Bedford, 1974

LIPSON, E.
The History of the Woollen and Worsted Industries.
London, 1921; reprint 1950

LORENZEN, E.
'Rokke med 2 tene'
Arv og Eje.
Copenhagen, 1969

McCALL, H.
Ireland and her Staple Manufacture.
Belfast, 1855

MACDONALD, D.
Fibres, spindles and spinning wheels.
Toronto, 1944

McEVOY, J.
Statistical Survey of Co. Tyrone.
Dublin, 1802

MacGEORGE, A.
Old Glasgow.
Glasgow, 1879

Madrid – Museo del Pueblo Español.
Catalogue

MANN, J. DE L.
'A document regarding Jersey spinning in the P.R.O.'
Textile History Vol. 4.
1973

MANTOUX, P.
The Industrial Revolution in the 18th century.
1st English edition, 1928

MINCOFF, E. & MARRIAGE, M. S.
Pillow Lace, a practical hand-book.
London, 1907; reprinted Chicheley 1972

MITCHELL, J. O.
Old Glasgow Essays.
Glasgow, 1905

MITCHELL, L.
The Wonderful Work of the Weaver.
Dublin, 1972

MORTIMER, J.
Cotton Spinning: the Story of the Spindle.
Manchester, 1895

MORTON, W. E. & WRAY, G. R.
An Introduction to the Study of Spinning.
London, 1962

MURPHY, W.
The Textile Industries.
London, 1910

NEEDHAM, J.
Science and Civilization in China.
Vol. 4, part 2, section 27. Cambridge, 1965

OPIE, I. & P.
The Classic Fairy Tales.
London, 1974

PALLISER, MRS. B.
History of Lace.
London, 1875

PATTERSON, R.
'The Story of Wool', 'From Fleece to Yarn'.
Bankfield Museum booklet.

PENNINGTON, D. & TAYLOR, M.
A pictorial guide to American Spinning Wheels.
Sabbathday Lake, Maine, 1975

PINTO, E. H.
Treen and other Wooden Bygones.
London, 1969
PLOT, R.
The Natural History of Oxfordshire.
Oxford, 1677
PLUMMER, A.
The Witney Blanket Industry.
London, 1934
POLOCZEK, F.
Slovenskè Ludové Piesne
Vol. III 1956. Vol. IV 1964, Bratislava.
POPE HUMPHREY, F.
The Queen at Balmoral.
London, 1893
POWER, E.
Medieval People.
London, 1924
POWER, E.
Wool Trade in English Medieval History.
Oxford 1941; reprint London 1969
PONTING, K. G.
'Sculptures and Paintings of Textile Processes
at Leiden'.
Textile History Vol. 5.
1974

RAMSAY, G. D.
*The Wiltshire Woollen Industry in the 16th
and 17th centuries.*
London, 1965
RATH, T.
'The Tewkesbury Hosiery Industry'
Textile History Vol. 7.
1976
REHTMAIER, P. J.
Chronicle of Brunswick-Lüneberg.
Brunswick, 1722
RETTICH, H. E. VON
Spinnradtypen.
Vienna, 1895
ROGERS, J. E. T.
History of Agriculture and Prices in England.
Vol. VI 1583–1702.
Oxford, 1887
ROLAND DE LA PLATIÈRE, J. M.
*Arts et Métiers. L'art du fabricant d'étoffes
en laines.*
Paris, 1780

SALZMANN, L. F.
English Industries in the Middle Ages.
London, 1964

SCHEUERMEIER, P.
Bauernwerk in Italien . . . Vol. II.
Erlenbach – Zurich, 1956
SCHELVEN, A. L. VAN
'Textile in Hont en Steen verbeeld'
Rayon Revue Vol. X. no. 6.
Arnhem
SCHWEIZER, W. F.
'Meister Jürgen, de uitvinder van det
spinnewiel?'
Textielhistorische Bijdragen no. 7.
Hengelo, 1965
SELLIN, T.
Pioneering in Penology.
Philadelphia, 1944
'Senex' (REID, R.)
Old Glasgow and its Environs.
Glasgow, 1864
SHARP, M.
A traveller's guide to the Saints of Europe.
London, 1964
SINCLAIR, J. (compiled by)
Statistical Accounts of Scotland 1791–99
SINGER, C. (ed.)
A History of Technology
'Spinning and weaving' Patterson, R.
Vol. II 1956; Vol. III 1957. Oxford
SUNG YING-HSING (translated E-Tuzen Sun &
Shiou-Chuan Sun)
Chinese Technology in the 17th century.
London, 1966

Textile Institute.
Identification of Textile Fibres.
Manchester, 1965
Textile Institute.
Textile Terms and Definitions.
Manchester, 1963
THIRSK, JOAN
'The Fantastical Folly of Fashion. The
English Stocking Knitting Industry 1500–
1700'
Textile History and Economic History.
Manchester, 1973
THOMPSON, G. B.
Spinning Wheels.
Ulster Museum Catalogue to the Horner
Collection.
Belfast, 1963
THORNTON, P.
'Italian Craftsmanship in Wood'.
Arte Illustrata no. 47.
1972

Tilburg Nederlands textielmuseum
*Het textielambacht in de schilderkunst van de
16e de 20e eeuw.* (catalogue)
TOMLINSON, C.
  *The Useful Arts and Manufactures of Great
  Britain.*
  London, 1861
TOSCHI, PAULO.
  Arte popolare italiana.
  London, 1959

Universal magazine, 1749
URE, ANDREW
  Cotton manufacture of Great Britain.
  London, 1836
USHER, A. P.
  A history of Mechanical Invention.
  Cambridge, Mass., revised 1954

VALLINHEIMO, V.
  Das Spinnen in Finnland.
  Helsinki, 1956
VICKERMAN, C.
  The Woollen Thread.
  London, 1881
VICKERMAN, C.
  Woollen Spinning.
  London, 1894

WADSWORTH, A. P. & MANN, J. DE L.
  *The Cotton Trade and Industrial Lancashire
  1600–1780.*
  Manchester, 1931; revised 1964
WALTON, MRS. O. F.
  Pictures and stories from Queen Victoria's life.
  London, 1893
WARBURG, L.
  Spindabog.
  Borgen, 1974
WARDEN, A. J.
  The Linen Trade Ancient and Modern.
  London, 1864
WEYNS, J.
  Volkshuisraad in Vlaanderen.
  Antwerp, 1974
WHITE, L.
  Medieval technology and social change.
  Oxford, 1962
WHITING, G.
  Old-time Tools and Toys of Needlework.
  New York, 1928; reprint 1971
WIKLUND, S. & DIURSON, V.
  Textil materiallära.
  Stockholm, 1962

YARRATON, A.
  England's Improvement by Sea and Land . . .
  London, 1677

ZELENIN, D. K.
  Russische (ostlavische) Volkskunde.
  Berlin, 1927
ZONCA, V.
  Novo Teatro di Macchine et Edifici.
  Padua, 1607

Source material for Appendices

ALDERSON, G. L. H.
  'The rare breeds survival trust livestock
  survey 1974'.
  (*The Ark,* May 1975)

British Sheep.
  Hand-book of the National Sheep
  Association, 1968
British Sheep Breeds, their wools and its uses.
  British Wool Marketing Board

HAIGH, H. & NEWTON, B. A.
  The Wools of Britain.
  London, 1952
HORNE, B.
  Fleece in Your Hands.
  New Zealand Spinning, Weaving & Wool-
  crafts Society Inc.

PONTING, K. G.
  'Wool quality'.
  (*The Ark,* June 1975)
PRIOR, M. & P.
  Wool and wool-sorting.
  Dryad Press. (No longer available.)

Books on practical spinning

ANDERSON, BERYL.
  Creative spinning, weaving and plant dyeing.
  New York, 1971
CASTINO, RUTH A.
  Spinning and dyeing the natural way.
  New York, 1974
CHANNING, MARION.
  The Magic of Spinning.
  Marion, 1966
DAVENPORT, ELSIE.
  Your Handspinning.
  London, 1953; reprint California, 1964
DUNCAN, MOLLY.
  Spin your own wool.
  London, 1973
FANNIN, ALLEN.
  Handspinning, Art & Technique.
  New York, 1970
GRASETT, K.
  Complete guide to hand spinning.
  London, 1930; reprint California, 1971
HOPPE, ELIZABETH, & EDBERG, RAGNAR.
  Carding, spinning, dyeing.
  New York, 1975 (translated from Swedish
  by Marianne Turner)
HORNE, BEVERLEY.
  Fleece in your hands.
  New Zealand Spinning, Weaving and
  Woolcrafts Society Inc.
KLUGER, MARILYN.
  The Joy of spinning.
  New York, 1971
LEADBEATER, ELIZA.
  Handspinning.
  London, 1976
Wool Textile Delegation.
  The Tex System.
  Bradford, 1975

Publications

Australia
The Australian Handweaver and Spinner,
The Handweavers and Spinners Guild of
 Australia,
Box 67 G.P.O.,
Sydney, 2001

British Isles
Crafts,
Crafts Council,
8 Waterloo Place,
London SW1Y 4AT

The Weavers Journal,
Association of Guilds of
 Weavers, Spinners & Dyers
BCM 963
London WC1N 3XX

Canada
Canada Crafts,
Page Publications Ltd,
380 Wellington Street, West,
Toronto M5V 1E3

New Zealand
The Web,
P.O. Box 225,
Balclutha,
Otago

U.S.A.
Craft Horizons,
American Craftsman's Council,
22 West 55th Street,
New York

Shuttle, Spindle and Dyepot,
998 Farmingham Avenue,
West Hartford,
CT. 06107

Index

Page numbers in *italic* type indicate plates, figures or their captions

Aberford, Lotherton Hall 172
accelerating spindle head 65–7, *67*
Act of Parliament in Scotland 122
Alexandra, Queen 174
Amsterdam 99, 117, 179
Angora rabbit hair, to spin 208, *227*
Antis, J. 165, 166, 167, 168
Arabs 27, 39
Arkwright, Richard 110, 123, *124*,
    152n, 177, 188, 190
Arles, Musée Arlatan 95
Ashford spinning wheel 81n
Aubrey, J. 181
Australia 30, 203n, 239
Austria (see also Tyrol):
    distaffs 95, 97;
    double flyer spinning wheel 154;
    flyer spinning wheels 128;
    flyers from 75, 76;
    reels 108;
    spinning wheel maker 142, 154

backing-off 43, 188
Bakewell, Robert 30
Baltic lands 97
band, open and closed *43*, 44 (see also
    driving, friction, *Kunkel*)
'barge' wheel 133n
Barnsley, Cannon Hall 170
Basel, Museum für Volkskunde 140
bat head 62
Bath, American Museum in Britain
    147
Bavaria:
    flyer spinning wheels 119, 129, *130,
    131*, 134, *138*, 158, 159;
    flyers from 75;
    reels 108, *109*
bearings:
    of flyer spinning wheel 73, 135;
    of Picardy wheel *113*, 114, 115;
    of spindle wheel *42*, 56
Beck's *Draper's Dictionary* 62
beetling 17
Belfast 172, 173;
    Ulster Museum 13, 115, 144, 172
Belgium 21, 149n (see also Flanders,
    Low Countries):
    distaff for plying 105, *105*;
    flyer spinning wheels *116*, 117, 128,
    133, 133n;
    Lys river 19;
    spindle wheels *42*, 56, 58;
    water pots 103
belt spinning wheel 162–4 (see also
    girdle spinning wheel)
'Bern' wheel 140
big wheel 60, *61*
Blaise, Bishop 35n
bleaching 17, 220–1
bobbin 16, 106;
    drag 80, *80*, 81n, *113*, 152n (see also
    friction band);
    holders 105, 106;
    lead 77–8, *77*, 90 (see also doubled
    band drive, flyer drag, friction
    band);

on flyer spinning wheels 76, 76–7,
    82, 85;
    winder (see under winders), lace
    maker's 162
bobbin' wheel 68
Bolivia 42
Bolton, Hall-i'th'Wood Museum 123,
    163
Bombyx mori 38
Book of Trades 32, 34, *36*
Borghesano 82
Born, W. 45, 60, 65, 92, 138
boudoir spinning wheel 158–60, 159n,
    160
bow:
    for cotton 26, *26*, 37;
    for silk waste 39;
    to make 209
breaker (for flax) 21–2, *22*
Brecknockshire Society 177
'bride's' spinning wheel 117
Brighton, Royal Pavilion 159n
Britain 75, 76, 132, 137, 143, 191 (see
    also England, Ireland, Scotland,
    Wales):
    distaffs 95;
    spindle wheels 54, 59–63, 154
British Wool Marketing Board 229
Bruegel, P. 95
Brunswick 91, 92:
    Landesmuseum 104, 133
Bulgaria 48
Burghley, Lord 90
Burma 47

cabling 218
cabriole legs 136, *137*
calculations 110, 110n, 151n
cam 152n, 165, 166, 167, 187, 188
Canada 12, 66, 67, *67*, 115
Canary Islands 68
Cannabis sativa 18
carders 31, 32, *33*, 35, 37, 88, 185, *185*,
    195–8, 205;
    to clean 198n
Cardiff (see St Fagans)
carding:
    cotton 185, 209;
    variations 223–4;
    with drum carder 206;
    wool 31, 37, *55*, 56, *57*, *146*, 195–8,
    195–8, 205, *205*
carding bench 31, *37*
carduus 35
'carriage' spinning wheel 158
castle spinning wheel 144–6;
    with double flyer 153, *157*
cat hair, to spin 208
Catalonia 24, 27, 53
Cecil, Anne 90
Cecil the spinner 94
Celle Museum 50
chair makers 149;
    Guild of 149n
chair spinning wheel 147–9, *148*
Channel Islands 61, 183
Chardin, S. J-B. 161

Charka 44, 45, *46*, 47, 49 (see also
    spindle wheel, rimless wheel):
    two driving wheels 65–6, *66*
Charlotte, Queen 170, 171
Chien Hsüan 45
China 18, 26, 39, 47–8, *47*
Chur 49:
    Rätisches Museum 154
cloth:
    finishing 16
    industry 180–2
    making of 15–6
clothier 180, 181
Cobbett, W. 189
Codde, Anna 88, 89, *89*, 90, 95, 101, 112
comb, flick 203n
combed wool 31, 32, 82, 95, 112
combing wool 31, 32–5, *35*, 187, 202–3,
    202:
    drawing-off 35, *36*
combs (for wool) 31, 32–5, *34*
cone 43, *43*, 58, 108;
    to build 222
cop 43, *43*:
    to build 222
Copenhagen 50, 97, *135*, 136, 153:
    Kunstindustrimuseet & Herning
    Museum 147
cotton 17, 18, 25–7, *26*, 41, 66, 161,
    221, *227*:
    bowing 26, *26*, 209;
    draw 209–10, *210*, 211, 222;
    early manufacture of 27;
    for knitting frame 184–5;
    ginning 25–6;
    mercerization 25;
    preparation for spinning 209;
    staple of 25, 209;
    spinners of 49, 62, 184, *185*, 186;
    spinning 41, *46*, 52, *52*, 180, 185,
    185, 209–10, *210*
'cotton' wheel 62–3, 184
Courbet, G. 103
crank 72, 193, *193*, 198
crank handle:
    on flyer spinning wheels 71, 72, 123,
    158, 161, 171;
    on spindle wheels 45, 49, 50, 56;
    on winders 68
creel *31*, 68
Crommelin, L. 99, 100. 115, 116, 117
Crompton, S. 123, *125*
croppers 16, 88
Czechoslovakia 95, 97, 132, 134

Dale, David 177
Decretals of Gregory IX 32, 54, *55*
Defoe, D. 84, 182
Delawny, Mrs 170, 170n
Denmark 50n:
    distaffs 97, *98*, 99, *99*;
    double flyer spinning wheel 153;
    flax breaker 22;
    flyer spinning wheels 120, 132, *135*,
    136, 136n;
    spindle wheels 50;
    two wheel spinning wheels 147;

wool combs 35
Dicziunari Rumantsch Grischun 49, 132n, 140
Diderot (*see Encyclopédie*)
Dipsacus fullonum 37n
Dipsacus sylvestris 37
disc wheel 45, 65, *66, 116,* 117n, 147
distaff 53, 70, *71,* 86, 94–9, *96,* 100, *118, 121,* 123, *126, 131,* 133, 142, 145, 161, 191, *211, 212*:
  bat-like 97, 101, *101;*
  comb *96,* 97, 100;
  cup-shaped 112, *114;*
  for plying 105, *106;*
  for wool 88, 89, 95, 112, 113, 183, 184, 205n;
  free-standing 86, *87,* 88, 95, *138,* 140;
  L-shaped 97;
  lantern 95, *124,* 163;
  position for spinning 213;
  to make 216n;
  tow 97, *98, 99,* 216;
  used with hand spindle 35, 40, 48, 94, 183
distaff dressing (*see* dressing the distaff)
diz 35
dog hair, to spin 208, *227*
Dou, Gerard 112, *114*
double flyer spinning wheel 149–58, *150, 153, 155,* 184
doubled band drive (bobbin lead) 77–8, *77,* 79n, 117, 119, 127, 135, 136, 138, 141, 142, 149, 150, 159, 166
Doughty, J. *166,* 167, 167n, 170
drafting 40, 89, 149, 188, 203–4, *204,* 214
drag (*see* bobbin drag, flyer drag, friction band)
Draperies:
  New 181, 183n, 184;
  Old 181
drawing-out 15, 40, 43, 54, 185, 188, 199–200, *199–200,* 222, 223
drawing-room spinning wheels 158–60, 161
dressing the distaff 99–101, *101,* 102, 136, 150, 211–3, *211, 212,* 216;
  eel skin used 102
driving band 42, 45, 48, *71,* 77, 78, 80, 81, 85, 88, 114, 121, 150, 151, 154, 158, 165
driving wheel (*see also* disc wheel, rimless wheel etc.):
  of flyer spinning wheel 70, *71,* 72;
  of spindle wheels 42, *42,* 44, 47;
  of winders 16, 68;
  position on castle wheel 144;
  positioned on the left 132n
drum carder 206, *206*
Dublin, National Museum of Ireland 157
'Dutch' wheel 116, 117, 121
dyeing 16, *31*
Dyer, J. 184

Edinburgh 152, 161, 171:
  National Museum of Antiquities of Scotland 157, 159
Encyclopédie 26, 69, 95, 100, 104n, 108, 123, 125, *126,* 127, 159, 161
Endrie, W. 48, 53
England 22, 27, 30, 37, 111, 178, 183n (*see also* Britain):
  Bridport twine spinners 63–4, *64–5;*
  Derby silk twisting mill 84;
  double flyer spinning wheels 149, 155;

early flyer spinning wheels 90;
East Anglia worsted industry 109, 181–2;
  flyer spinning wheels 120, 122–3, 135, 138
  knitting industry in 183;
  lace-making in 100, 178n;
  Lake District 174, 175, 190, 191;
  Plymouth sail canvas manufactory 154–5;
  revival of flax spinning in 190–1;
  seats for spinning 104;
  Shrewsbury clock-makers 163;
  spindle wheels 60, 62, 182;
  spinners in 62, 183, 184, 185;
  spinning wheel makers in 159, 163, 167, 168;
  West Country woollen industry 180–1;
  Witney blanket industry 182n;
  Yarmouth twine spinners 64;
  Yorkshire cloth industry 182
Evelyn, J. 179

Faeroe Islands 58, *59,* 59n
fairy tales 127n, 175
Farnham Museum 115, 158
Feather, T. 68
Feldhaus, F. M. 92
felloes 70, 123, 125, 138
fibres 15, 17, 45, 48, *227;*
  animal 17;
  bast 17, 48;
  factors when choosing 226;
  leaf 17;
  man-made 17–18, 208;
  mineral 17;
  seed 17
Fiennes, Celia 178n, 184
Finland 97, 101, 120
Firmin, T. 149, 150, *150,* 151, 170n, 178
Flanders 27, 30 (*see also* Belgium, Low Countries):
  early flyer spinning wheel 90;
  early use of spindle wheel 54;
  flyer spinning wheel 117;
  spinning fine flax in 99, 100
'Flanders' wheel 54
flax 17, 18, *20,* 18–25, 63, 82, 95, 99, 115, 150, 190, *227* (*see also* distaff, dressing the distaff, linen, ribbons & bands, etc.):
  breaking 21–2, *23;*
  dressing 21, *23;*
  hackling 21, *23,* 24, 211;
  line 24, 211;
  on early flyer spinning wheels 85, 86, 90;
  preparation for spinning 211;
  retting 19–21, 19n, 20n, *21;*
  rippling 19;
  scutching 21, 23, *23;*
  tow 24, *98,* 143, 150
flax spinning *98, 102, 103, 116, 126,* 127, 133, *135, 138,* 140, 154, 167, *174,* 213–6, *214:*
  according to Crommelin, 99–100
  Encyclopédie, 127
  Firmin, 151
  Garnett 191;
  dry 103, 214;
  fashionable 136, 158;
  fine 100;
  from the finger 191, 224;
  revival of 190–1
flax spinning wheel 69, 113, 116, 136–7, 146, 179

fleece 28:
  choosing 229–30;
  lay-out for sorting *231;*
  rolling 29;
  spinning straight from 208;
  weight 30, 233–39
fluted pillars 138
flyer 69, 73, *76,* 125, 140, 149:
  drag 78–9, *79,* 79n, 84, 128, 129, *131, 132,* 134, 136, 140, 158, 190;
  holes in 76, *79,* 136, 140;
  hooks on 75;
  in silk throwing machine *83,* 84;
  lead (*see* bobbin drag);
  mechanism *71,* 73, 74–6, *76,* 77, 144;
  on Picardy wheel *113, 114,* 115
flyer spinning wheel (*see also* double flyer, horizontal, vertical etc.):
  general description 69–82, *71;*
  early evidence in Europe 82–92, *86, 87,* 89
folklore 175–6
footman *71,* 72–3
France 21, 35, 53:
  distaffs 95, 99, *126;*
  double flyer spinning wheel 153;
  early evidence of spindle wheels 53, 54;
  early flyer spinning wheels 90;
  flyer spinning wheels 125–8, *126, 129,* 134, 136, 141–2;
  hank holders 111, *126;*
  Picardy wheel *113, 114;*
  reel 108, *126;*
  ribbon for flax 101, 102;
  spindle wheels 54, 56;
  spinning 114
friction band:
  for bobbin drag 80, *80,* 81, 85, 88;
  for flyer drag 79, *79;*
  on Picardy wheel *113,* 115
fustian:
  cloth 27, 184;
  weavers 53

Galle, van J. 88
Gandhi, M. K. *66,* 66
Garnett, A. 191, 213
'German' spinning wheel *79,* 128, 159
Germany 27, 35, 37, 117 (*see also* Bavaria, Saxony, Lower):
  distaffs 94, 95, 97;
  double flyer spinning wheel 154, *155;*
  early flyer spinning wheels 85, 90;
  early record of wheel spinning 53;
  flax breaker *22;*
  flax dressing 21, 23, 24;
  flax spinning school 178;
  flyers 76;
  flyer drag 79, *79,* 128;
  flyer spinning wheels 117, 119, 129, 134, 136, 140, 164;
  girdle spinning wheel 163–4;
  hank holders 111;
  Kunkelband 101–2;
  reels 108, 109;
  ribbons for flax 101;
  spindle wheels 50, 53;
  treadle on vertical wheel 92;
  water pots 103;
  work houses in 179–80
girdle spinning wheel *14,* 162–4, *164*
Glasgow 177:
  Art Gallery and Museum 122
Glockendon Bible 85, 92
Gloucester 184:
  City Museum and Art Gallery 155

'goat' wheel 49
'gossip' wheel 157
Gossypium 25
Gray, A. 93, 152, 166
grease pots 104
great wheel 59, *59, 60,* 62, *63,* 108, 113, 181, 183, 184
Greece 35, 38, 48, 68
Guest, R. 185, *185*

Habib, I. 44, 45
hackles 24:
  substitute for 211
hair (of animals) 17
Hall, R. 21, 24, 102, 116, 177
Hamilton, Lady 123
hank holders 110–1, *111, 126*
Hanover 117, *118*:
  Historisches Museum 119
Hardy, J. 167, 168
Hargreaves, J. 38, 187, 187n, 188, 190
Heemskerck, Maerten van 88, *89*
hemp 17, 18, 63:
  for twine spinning 63–5, *64–5;*
  spinning wheel 113, 117n
Herrick, R. 176
'high' wheel 50
Hilleström, P. 24, 52, *52,* 53
Hincks, W. 23, *23,* 24, 102, *103,* 108, 121
Hoffmann, M. 50, 59n, 152, 154
Holland 21, 108, 122, 181n, 213 (*see also* Low Countries):
  bobbin holder 105;
  carding bench *31,* 37;
  distaffs 97, 105, *105;*
  double flyer spinning wheel 154;
  flax dressing 21, 24;
  flyer spinning wheels 116, 117, 133n;
  hemp spinning 65;
  rippler 19;
  spin house 179;
  twisting wheel 117
Honourable Board of Trustees for
  Manufacturers 151–2, 152n
hoop rim 117:
  absence in Far East 54;
  early flyer spinning wheels with 89;
  flyer spinning wheels with 123, 125, *125,* 138;
  making 58;
  Picardy wheel with 112, *114;*
  spindle wheels with 48, 49, 53–64
horizontal spinning wheel 92:
  fashionable 159;
  with frame 125–32, *126, 130, 131,* 132n;
  with stock 115–25, *71, 118, 119, 121,* 132n, 144, 168, *173,* 191;
  with triangular base 128, *129*
Horner Collection 13, 68, 108–9, 115, 117, 119, 132, 138, 140, 143, 154
Horner, J. 13, 48, 60, 64, 99, 100, 128, 132, 154, 213
hosiers 53, 185
houses of correction 50, 50n, 179, 180
Howard, J. 50n, 179, 180
Huddersfield broad wheel 123
Hungary 132, 134, 147
'hurdy' wheel 67, *67*

Iceland 59
'improved' spinning wheel 164–70, *166, 169*
India 18, 44, 56, 58:
  charka 44, 45, *46*
  charka, two wheel 65

cotton bowing 26
cotton gin 25
hand spindle 41
spindle wheels 45
Indonesia 47
Industrial Revolution 184, 189
inventories 90, 171, 184
Ireland *20, 21,* 102, *103,* 170n, 190 (*see also* Statistical Surveys of Monaghan and Tyrone):
  Aran islands 60;
  big wheel 60, *61;*
  bleaching linen 17;
  bruising flax 22–3;
  castle wheel 144–6, *146;*
  distaff dressing 99–100, 213;
  double flyer spinning wheel 151, 157;
  'Dutch' wheels in 116, 121;
  flax dressing 22, *23,* 24
  multiple spinning wheel 156, 157
  spinners in 176;
  spinning in 60, 186, 192;
  spinning wheel maker 172–3
Isle of Man 54, 104, *111,* 175, 190
Italy 75:
  cotton manufacture in 27;
  distaffs 95, 97;
  early flyer spinning wheel 84;
  flyer spinning wheels 132, 134, 142, 143;
  hackle boards 24;
  reels 108;
  silk in 39;
  silk throwing machines in 82;
  spindle wheels 49;
  water pots 104;
  winders 68;
  wool manufacture in 30

James, J. 62, 182, 183
Jameson, J. 159, *160*
jersey yarn 61–2, 183, 187
Jersey wheel 61–2
Jonge de, J. R. *178*
Jurgen, J. 91–2

Kay, J. 187
Keighley, Cliff Castle Museum 68, 123
knob turnery 123
Krakow, Ethnographic Museum 142
Kunkelband 102, 115, 163, 164

lace 17, 100, 178n, 220, *228*
'Laken' *31*
Lane, Mrs *14,* 163
Lapland 96
Lazy Kate 105, *106,* 217, 217n, 222, 225, *226*
Leake, J. 183n
Lee, William 184
Leeds, Temple Newsam House 170
Leyden, L. van 87, *87,* 88
linen 16, 17, 85, 220:
  half-bleaching 221;
  measuring table 110;
  tying skeins of 219, *219*
  to soften 220
  yarn, 179n, 220n
'linen' wheel 90
Linum angustifolium 18
Linum usitatissimum 18
Livre des Mestiers (le) 94
Lombe, J. 84
London 149, 159, 178, 184:
  Bridewell 179;
  National Gallery 68;

Science Museum 48, 110, 123, 154, 162, 170;
  Victoria and Albert Museum 159, 161
Londonderry, Lady 160
long draw 198, 199–201, *199–200,* 209n, 210, 218, 222
long fibre wheel 69
'long' wheel 60
loom 15, 16
'lover's' wheel 157
Low Countries (*see also* Belgium, Flanders, Holland):
  carding bench 38;
  distaffs 95, 97, *106;*
  early flyer spinning wheel 86, 88, 90;
  flyer spinning wheels 134;
  'Kunkelband' 101
Lucknow painting *46*
Lutrell Psalter 54, *55,* 85, 186

machines:
  carding 188;
  gig-mill 37n;
  jenny 187–8, 189;
  knitting frame 184;
  mule 123, 186;
  ring frame 189–90;
  roller drafting 188;
  saw gin 26;
  silk throwing 82–4, *83;*
  slubbing billy 186, 188;
  spinning 184, 186, 187;
  spinning frame 188
Madrid 113:
  Museo del Pueblo Español 135
maidens 71, 73, 112
Marshall, Mrs M. 167, 167n
McCall, H. 121, 157, 160, 170n, 177
McCreery, J. & son 172, 173, *173,* 174, *174*
Millet, J. F. 128
Minor, A. 66, 67
Mittelalterliches Hausbuch 85, 86, *86,* 91
mohair, to spin 208, *227*
'mother and daughter' wheel 157
mother-of-all 71, 73, 123
'Mother-of-God' wheel 49
muckle wheel 60
Munich:
  Bayerisches Nationalmuseum 158
  Deutsches Museum 50, 154, 164
Museums (*see under* place names)

Navajo:
  plying technique 225, *225*
  spindle 41
Needham, J. 44, 45, 47
Newbury, Fr. H. *64*
New Zealand 81n, 203n, 239
niddy-noddy *102,* 106, *126,* 219, *219*
noils 32, 34
'Norrawa' spinning wheel 122, 137
Norrbotten Museum 142n
North America 184, 189 (*see also* Canada, United States):
  accelerating spindle head 66;
  castle spinning wheel 146;
  Cobbett's remarks 189;
  Connecticut chair spinning wheel 147–9, *148;*
  distaffs 95;
  double flyer spinning wheel 157;
  two wheel flyer spinning wheels 147–9
  wool wheel 60
Norway 50:
  disc wheel 117n;

distaffs 95, 97;
double flyer spinning wheel 152;
'English' spinning in 52;
flyer spinning wheels 120, *121*, 136;
reel 108
spindle wheels 50–1
Norwich 32, 109, 181, 182:
Stranger's Hall 105n

oil for wool 30, 176, 194
one thread wheel 42, 182, 189
Oriental-type spindle wheels 48–52, 53,
56 (*see also* rimless wheels, spindle
wheels)
orifice 74, *76*, 114, 117, 161, 167
Oslo 50:
Norsk Museum 51, 117n
Oxford, Pitt Rivers Museum 155, 159,
163

Pakistan 45
Palliser, Mrs 100
Paris 53, 99:
Musée des Arts et Metiers 134
Musée National des Arts et Traditions
Populaires 142
Paul, Lewis 62, 184, 186, 188
Persia 45
Peru 18
Picardy spinning wheel 112–5, *114*, 183
Picardy-type flyer 112, *113*, 117
picking (wool) 37 (*see also* teasing)
Pintoricchio of Perugia 68
Planta, John 168, *169*, 170
Plot, Dr R. 182n
plying 16, 39, 88, 105, 159, 161, 163,
216–8, *217*, 225, *225*, *226*:
direction of 76, 217, 225;
distaff for 105, *105*;
spinning wheels used for 117, 133,
140, 142n
with spindle wheel 44, 67, 222
Poland 119, 142
Portugal 21
Prestwich, Heaton Hall 170

quality count 29, 233–9

Rare Breeds Survival Trust 229
reels *102*, *103*, 106–110, *107*, *108*, *109*,
126, 171, 172
reeling *103*, 106–8, *126*, 218–9, *219*, 222
Rehtmaier, P. J. 91
Rettich, H. E. von 93, 99, 128, 142, 147,
151n, 152
ribbons 101–2, 213
rice 111, *111*, 217
rimless wheels:
in the East 45, *46*
in Europe 48, 49, 50, *51*, 53
rippler 19
rock 91, 94, 181 (*see* distaff)
Rogman, G. 102, *102*
rolag 37, 42, *46*, *52*, 54, 56, 62, 186,
197, 198, *198*, *199–200*, 206, 209,
221, 223, *227*
Roland de la Platière, J. M. 45n, 113,
114
Romania 134
Rome, Museo Nazionale delle Arti e
delle Tradizioni Popolari 49
Romney, G. 123
roving *47*, 48, 94, 97, 112, 113, 161,
185, 186, 187, 203, 206, 209
Ruskin, J. 190, 191
Ruskin Linen Industry of Keswick 191
Russia 97, 119

S twist *43*, 44, 44n, 54, 76, 82, 85, 140,
146, 205, 213, 223, 225
St Aubin de, G. 153
St Distaff's day 176
St Fagans, Welsh Folk Museum 125, 137
St Katherine 176n
Sandby, Paul *14*
Saxony, Lower 85–6, 91:
distaff 97, *118*;
distaff dressing 100;
double flyer spinning wheel 154;
flyer spinning wheels 117, *118*
spinning chair 104
'Saxony' wheel 91, 173
Scandinavia 12, 111, 122, 213 (*see also*
Denmark, Finland, Norway,
Sweden):
distaffs 97;
flyer spinning wheels 120, 132, 134;
'*Kunkelband*' 101
spindle wheels 50
Scheuermeier, P. 48, 49, 132n, 142, 143
Schwarz, A. 49, 53, 58
Schweizer, W. F. 91
Scotland 24, 29, 30, 35, 64, 104, 110,
122, 136, 151, 159, 171, 175, 177 (*see
also* Statistical Account):
Berbecula, island of 122;
bobbin drag in 81, 125n;
castle spinning wheel 144, *145*;
distaff 97, *96*;
double flyer spinning wheels 151–2,
153
flyer spinning wheels *96*, 122, 133,
134, 135, 136, 137, *137*, 161, *162*;
grease pots 104;
Harris, isle of 122, 192;
Hebrides 179n;
Highlands of 179;
Mull, isle of 174;
St Kilda, island of 122;
seats for spinning 104;
Shetland isles *80*, 122, 137, 175, 191;
Skye, island of 192;
spindle wheels 56, 58, 60;
spinning in 191–2;
spinning schools 179;
spinning wheel makers 151, 152, 157,
160, 161, *162*, 171, 172, 173, 174;
three flyer spinning wheel *156*, 157
scouring 16:
linen 220
wool 30, *31*, 32, 194, 220
seats 104, 170 (*see also* spinning chair)
semi-worsted yarn 207
sericulture 38
shears 28, 29, *29*
sheep 30:
primitive 27, 30;
shearing *28*, 29, *29*, 229
sheep breeds:
British 30, 229, 233–9;
cross-breds 229, 239;
four-horned 235;
long wools 229, 236–7;
merino 29, 30, 181, 229, 233, 239;
mountain 237–9;
New Zealand 239;
Scottish island 235;
short wools 229, 233–5
Sheraton style spinning wheel 168, *169*
short draw 205–8, *207*, 223
short fibre wheel 42
shuttle 16, 17
Sicily 39, 68
silk 17, 18, 38–9:
reeling 38, 82;

throwing 39, 44, 82;
throwing machines 82, *83*, 84;
twisting wheel 49, 105n
silk waste 18, 39, 141n:
spinning 210–1, *227*;
spinning wheels used for 141, 161
size 16
slivers 35, 209
slubbings 186, 187, 188, 189
slubs 41, 224, *228*
Snowshill Manor 62, 138, 155, 163n,
170
Society of Arts 181, 187:
Secretary of 165, 181;
Transactions for 165
Spain 27, 39, 53, 113, 135 (*see also*
Catalonia)
Spin House *178*, 179
spindle:
hand *35*, 40–2, *41*, 58, 64, 94, *126*,
183–4, 222–3;
of bobbin winders 16, 44, 68;
of flyer spinning wheel 74, *76*, 88;
of spindle wheel *42*, *43*, 48, 51, 58,
222
spindle wheels 42–67 (*see also* Charka,
great wheel, rimless wheels, wool
wheel etc.):
early evidence 45, 53, 54, *55*;
general description 42–3, *42*;
in British Isles 59–64, *60*, *61*, 63;
in Europe 48–65, *51*, *52*;
in the East 44–8, *46*, *47*;
in woollen industry 56, 57;
inspiration for machine 187;
multi-spindle 47, *47*;
spinning on 42–3, 54, 56, 221–2, *221*;
with hoop rim 53–9
spinners:
independent 183;
in industry 57, 180;
in Nottingham 184–5;
lady 52, 90, 123, 137, 160, 163, 165,
170, 177;
market 180;
of candle wicks 184;
of cotton 186;
of fine wool yarns 181–2;
of Norfolk 183–4;
of twine 63–4;
servant 176–7;
with failing sight 170n
'spinnie' 137
spinning 13, 15, 39, 45n, 177 (*see also*
cotton, flax, wool, woollen,
worsted):
by children 61, 150–1, 179, 180 181;
by men *31*, 64, 181;
chair 104, *105*, 170;
competitions 177;
schools 150–1, 178–9;
songs 175–6;
specialized trade 180;
with bobbin drag 81;
with early flyer spinning wheel *87*,
89, *89*;
with hand spindle 40–1, 222–3;
with spindle wheel *31*, 42–3, 49, 53,
55, *57*, 183, 186, 221–2
spinning wheel 16 (*see* flyer spinning
wheel, spindle wheel etc.):
cost of 59n;
decoration of 62, 119, 133–4, 158,
161, 168, 172;
earliest form 42;
for two people 157;
improvements to 151–2, 152n, 164;

in work houses 180;
inspiration for machine 188;
lubricating the 104, 194;
multiple *156*, 157;
when buying 78n;
with musical box 164;
with reel attached 108–9;
with three flyers *156*, 157;
woods used in making 69, 122, 128, 158, 167, 168, 172
spinning wheel makers 12, 122, 116, 149 (*see also* Austria, England, Ireland, Scotland and individual's names)
spool 16, *31*, 68:
  rack 217
staple length:
  of cotton 25, 209;
  of wool 27, 194, 202, *227*, 233–9
starter thread 82, 194, 198, 214, 217, 218, 221
Statistical Accounts of Scotland 20, 99n, 151, 176
Statistical Survey of Co. Monaghan 22, 24
Statistical Survey of Co. Tyrone 176
Statutes of the Drapers of Chauney 32
stays 72, 113, 115, 120, *121*
Stewart, P. 171, 172, 173, 174
stock 70, *71*, 87, 88, 104, 120, 172:
  fiddle shaped 122
stock cards 37, 38
Stockholm 50:
  Nordiska Museum 144, 152, 153, 157, 164
Strasbourg calendar 86
strusa 39:
  to prepare and spin 210–1
Swanenburgh van, I. N. 112
Sweden 50, 102:
  bobbin winder 136;
  cotton spinning in 52, *52*;
  distaffs 97;
  double flyer spinning wheel 152, 153;
  flyer spinning wheels 120, 133, 136, 164;
  girdle spinning wheel 164;
  hackling 24;
  plying wheel 142n;
  rippler 19;
  spindle wheels 50, 52, *52*, 53, 58
swift 16, 48, 68, 110, 111, *111*, 217
'Swiss' wheel 140
Switzerland:
  cotton spinners in 49;
  distaffs 95, 97;
  double flyer spinning wheel 154;
  flyer spinning wheels 119, *119*, 129, 134, 140–1, *141*;
  Kunkelband 101;
  ribbon for flax 102;
  rimless wheels 49, *51*;
  spinning silk waste 141, 141n

table spinning wheel 161–2, *162*
teasing (wool) 37, 195, 202, 206, 208
teasles 16, 35, 37, 37n, 88
tensioner:
  on flyer spinning wheel *71*, 73–4, *74*, 123, 125, 134, 135, 140, 149;
  on spindle wheel 56, 57, 61
tensioning 78, 88–9, 194
Thessaloniki, Ethnological Museum of Macedonia 48
thistle 35, 37
thread 15 (*see also* yarn):

breaking when reeling 165, 215;
  for sewing 17, 190
threading hook 70, 82, 173
Tintoretto, J. 49
tops 35, 112:
  to spin 183, 208–9, 210
tow (*see also* distaff, flax):
  to spin 216
toy spinning wheel 158–60
trade cards 159, 172
trade directories 171, 172
treadle:
  first in evidence 92;
  in China 47–8, *47*;
  on flyer spinning wheel *71*, 72–3, 92–3, 128, 133, 158;
  two on a spinning wheel 147, 149
treadling 193
twining wheel 63
twist 15, 40–1, 43, *43*, 44, 44n, 53, 78:
  direction of in plying 76, 217, 225;
  in cabling 218;
  of silk 38–9
twisting 44, 49, 88, 140, 161:
  specialized trade 117, 184
twisting wheel 117, 105n
two handed spinning wheel (*see* double flyer spinning wheel)
two-wheel spinning wheels 147–9 (*see also* chair spinning wheel)
Tyrol:
  flyer spinning wheels 128, *131*, 134, 136n, 142;
  spinning chair 104, *105*;
  'Tyrolese' spinning wheel 128

United States of America 12, 26, 147 (*see also* North America)
Universal magazine 30, 108

Velasquez, D. 112, 113, 115
vertical spinning wheels 92, 132–43:
  with base *96*, 104, 105n, *131*, 132–138, 133n, *135*, 167, 174;
  with frame 138–43, *138*, *141*
Victoria, Queen 29, 171–74, *174*, 179n
Vinci, L. da 84–5
Vliet, J. J. van 149n

Waldstein Woollen Mill 57
Wales 59n, 176:
  carding bench 37;
  competitions 177–8;
  flax grown in 137;
  flyer spinning wheels 125, 137–8;
  spindle wheels 59, *60*
'walking' wheel 61
warp 15, 16, 30, 32, 53, 56, 68, 94, 177, 188, 217
warping mill 16, *31*, 68
washing (*see* scouring)
water-pots *96*, 103–4, 168
Watson, R. 172
Watts, R. 38, 68
weaving 15–6
Webster, J. 163, 164
Webster, R. 163, 163n
weft 16, 32, 39, 94, 188 (woof 30, 177)
'Welsh' wheel 125
Weyns, J. 90, 149n
wheel uprights *71*, 72, 120, 134
wheelwright (*see* spinning wheel makers)
Whitney, Eli 26
whorl:
  on flyer spinning wheel 69, 74–5, 76, 76, 77, 78, 78n, 80, 81

on hand spindle 40
on spindle wheel 42, *42*, 44, 48, 58, 63, *63*, 66
Williamsburg (Virginia), Wythe House 170
willowing 30, *31*
winder 16, 44, 48, 50, 67–8, 136, 170
winding 16, *31*, 49, 68, 163n
wool 17, 18, 27–38, 179, 182:
  Botany 30;
  choosing 229;
  combing (*see under* combing)
  crimp 27, 89, 230, 233;
  felting 27;
  for woollen spinning 194;
  for worsted spinning 202;
  from slaughtered sheep 28, 182n;
  grades 29, 229;
  in Middle Ages 30;
  lustre 27, 230;
  oiling 30, 32, 194, 207;
  quality of 29, 299–300;
  Saxony 179;
  scouring 30, *31*, 32, 194, 220;
  shrinkage 27;
  sizing 16;
  sorting of 29, 30, 230–1, *231*;
  Spanish 181, 184;
  Spelsau 205n;
  staple of 27, 32, *33*, 230;
  storing 231
  tying hanks of 219, *219*
  winder 111
  yarn count systems for 110
  yolk in 28
wool comber 32, 35, 35n, *36*
wool spinning 27, 28, *130*, *131*, 146, *146*, 181, 183
  short draw 205–9
  variations in 224–5
  with early flyer spinning wheel 87, 88, 89, *89*, 90
  with spindle wheel 54, 186
wool wheel 60, 66, 90
woollen:
  cloth 16, *31*, 228;
  spinning 55, 57, 120, 181, 194–201, *199–200*, 221–2;
  yarn 32, 183, 220, *227*
Wordsworth, W. 175, 189, 190
work houses 179
worsted:
  reel for 109;
  spinning 50n, 202–5, *204*, 236;
  yarn 31, 62, 179n, 220, *227*
worsted-type:
  spinning 112, 115, 183;
  yarn 32, 88, 89, 150, 184
'worsted' wheel 179

yarn 15, 17, *227* (*see also* thread):
  calculations 110–11;
  counts 29, 110;
  crepe 39;
  dealers 177, 180;
  design 226;
  for crochet 17, 216, 220;
  for knitting 17, 183, 216;
  unevenness in 177, 182
Yarraton, A. 24–5, 117, 178
York 159, 167:
  Castle Museum 167, 168
Yugoslavia 48, 133

Z twist *43*, 44, 44n, 75, 82, 198, 203, 223, 225
Zonca, V. 82, *83*